Wireless Networks

Series Editor

Xuemin Sherman Shen, University of Waterloo, Waterloo, Canada

The purpose of Springer's Wireless Networks book series is to establish the state of the art and set the course for future research and development in wireless communication networks. The scope of this series includes not only all aspects of wireless networks (including cellular networks, WiFi, sensor networks, and vehicular networks), but related areas such as cloud computing and big data. The series serves as a central source of references for wireless networks research and development. It aims to publish thorough and cohesive overviews on specific topics in wireless networks, as well as works that are larger in scope than survey articles and that contain more detailed background information. The series also provides coverage of advanced and timely topics worthy of monographs, contributed volumes, textbooks and handbooks.

Vimal Bhatia · Zhiguo Ding · Keshav Singh ·
Amit Baghel · Abhinav Singh Parihar ·
Deepak Kumar

Advanced NOMA Techniques for Heterogeneous Cellular Networks

Springer

Vimal Bhatia
Department of Electrical Engineering
Indian Institute of Technology Indore
Indore, Madhya Pradesh, India

Keshav Singh
National Sun Yat-sen University
Kaohsiung, Taiwan

Abhinav Singh Parihar
Department of Electronics
and Communication Engineering
Indian Institute of Information Technology
Bhopal
Bhopal, Madhya Pradesh, India

Zhiguo Ding
Department of Computer Communication
Engineering
Khalifa University
Abu Dhabi, United Arab Emirates

Amit Baghel
Department of Electrical Engineering
Indian Institute of Technology Indore
Indore, Madhya Pradesh, India

Deepak Kumar
Department of Computer and Information
Engineering
Khalifa University
Abu Dhabi, United Arab Emirates

ISSN 2366-1186 ISSN 2366-1445 (electronic)
Wireless Networks
ISBN 978-3-032-09550-3 ISBN 978-3-032-09551-0 (eBook)
https://doi.org/10.1007/978-3-032-09551-0

This Springer imprint is published by the registered company Springer Nature Switzerland AG
The registered company address is: Gewerbestrasse 11, 6330 Cham, Switzerland

Preface

The evolution of wireless communication systems has led to the development of advanced network paradigms to accommodate the ever-increasing demand for high data rates and enhanced connectivity. This book delves into the intricacies of non-orthogonal multiple access (NOMA) technology within heterogeneous cellular networks (HCNs), emphasizing the critical challenges and innovative solutions associated with hardware imperfections and advanced communication techniques. The discussion begins with an introduction to the significance of NOMA in next-generation wireless networks, providing an overview of HCNs and their role in improving spectral efficiency and user connectivity. Key challenges such as hardware imperfections and the importance of successive interference cancellation (SIC) are also explored, laying the foundation for subsequent analyses. Addressing the impact of high-power amplifier (HPA) non-linearity and imperfect SIC in NOMA-enabled HCNs, the book examines how these non-idealities influence network performance while demonstrating NOMA's resilience compared to orthogonal multiple access (OMA) techniques under practical constraints. A stochastic geometry-based approach is employed to analyze NOMA-empowered HCNs, considering the effects of imperfect SIC and channel state information (CSI). Insights into the spatial distribution of users and base stations using the Poisson point process provide a comprehensive understanding of network behavior under realistic deployment conditions. Further extending the analysis, the integration of simultaneous wireless information and power transfer (SWIPT) with cooperative NOMA is investigated, demonstrating its potential to enhance energy efficiency and user fairness. The role of interference is also reevaluated, particularly in smart homes, grids, and industrial IoT applications, where impulsive electromagnetic interference challenges traditional Gaussian noise assumptions. By leveraging stochastic geometry, the interplay between impulse noise characteristics and system performance is examined, revealing the advantages of NOMA over OMA even in impulsive noise environments. Beyond interference management, the book explores the potential of reconfigurable intelligent surfaces (RIS) in augmenting NOMA networks, focusing on their ability to enhance signal strength, coverage, and overall network performance. The concluding discussions summarize the key findings, highlighting advancements in addressing hardware imperfections and optimizing NOMA for HCNs. Future research directions are also outlined, including emerging technologies such as RIS and further refinements in NOMA-based networks. Serving as a comprehensive resource, this book offers researchers, engineers, and professionals valuable insights into the challenges and solutions associated with NOMA in heterogeneous cellular networks, bridging theoretical foundations with practical implementations to drive advancements in wireless communications.

Indore, India — Vimal Bhatia
Abu Dhabi, United Arab Emirates — Zhiguo Ding
Kaohsiung, Taiwan — Keshav Singh
Indore, India — Amit Baghel
Bhopal, India — Abhinav Singh Parihar
Abu Dhabi, United Arab Emirates — Deepak Kumar

Acknowledgements The completion of this book would not have been possible without the generous support and funding from esteemed institutions and research initiatives. We extend our sincere gratitude to the Ministry of Education, India, for funding the SPARC project, Connecting Billions of Devices in Beyond 5G Era, in collaboration with the UK, which significantly contributed to our understanding of next-generation networks. We also acknowledge the Integrated Optical and Radio Frequency Sensing and Communication for Sustainable Development project, which provided valuable insights into ISAC applications.

Special thanks to the Erasmus+ program and the University of West, Sweden, for supporting research on Mobility of Higher Education: M2M with 6G Communications. We are also grateful to the Technology Innovation Hub COMET at IIIT Bangalore for funding the Smart Radio Environment: Development of Controller for RIS. Finally, we appreciate the Ministry of Electronics and Information Technology, India, for supporting our work on Design and Implementation of Non-Orthogonal Multiple Access Techniques for Future Wireless Systems.

Competing Interests The authors have no competing interests to declare that are relevant to the content of this manuscript.

Contents

Abbreviations

3GPP	Third-generation partnership project
4G	Fourth generation
5G	Fifth generation
6G	Sixth generation
AF	Amplify-and-forward
AM/AM	Amplitude-to-amplitude
ASER	Average symbol error rate
AWGN	Additive white Gaussian noise
BS	Base station
BW	Bandwidth
CCU	Cell-center user
CDF	Cumulative distribution function
CDMA	Code division multiple access
CEU	Cell-edge user
CoMP	Coordinate multi-point
CRAN	Cloud radio access network
CSI	Channel state information
DF	Decode-and-forward
EH	Energy harvesting
EHS	EH using the superimposed signal
EHSI	EH using the superimposed signal and the aggregate co-tier and cross-tier interference
eMBB	Enhanced mobile broadband
FBS	Femto base station
FD	Full-duplex
FU	Femto user
Gbps	Gigabits per second
HCN	Heterogeneous cellular network
HCPP	Hard core point process
HD	Half-duplex
HPA	High-power amplifiers

i.i.d.	Independent and identically distributed
i-CSI	Imperfect channel state information
IEEE	Institute of electrical and electronics engineers
Im-N	Impulsive noise
IoT	Internet-of-things
IRS	Intelligent reflecting surface
i-SIC	Imperfect successive interference cancellation
LANs	Local area networks
LB	Lower bound
LoS	Line-of-sight
LTE	Long-term evolution
LTE-A	Long-term evolution-Advanced
MAC	Medium access channel
Mbps	Megabits per second
MBS	Macro base station
MIMO	Multiple-input and multiple-output
MMSE	Minimum mean square error
mMTC	Massive machine-type communication
mm-Wave	Millimeter-wave
MRC	Maximal ratio combining
MUST	Multi-user superposition transmission
NLD	Non-linear distortion
NOMA	Non-orthogonal multiple access
OFDMA	Orthogonal frequency division multiple access
OMA	Orthogonal multiple access
PA	Power amplifier
PB	Power beacon
PBS	Pico base station
PCP	Poisson cluster process
PDF	Probability density function
PF	Proportional fairness
PPP	Poisson point process
PS	Power splitting
QoS	Quality of service
RB	Resource block
RF	Radio frequency
RIS	Reconfigurable intelligent surface
RV	Random variable
SC	Selection combining
SEP	Symbol error probability
SIC	Successive interference cancellation
SINR	Signal-to-interference and noise ratio
SNR	Signal-to-noise ratio
SWIPT	Simultaneous wireless information and power transfer
TDMA	Time division multiple access

TS	Time switching
UB	Upper bound
UE	User equipment
URLLC	Ultra-reliable and low-latency communication
V-BLAST	Vertical-bell laboratories layered space-time
WPCN	Wireless powered communication network
WPT	Wireless power transfer

Symbols

- Basic arithmetic and calculus notations with their definitions.

Elementary and Special Functions

$\Gamma(x)$	$\int_0^\infty t^{x-1}e^{-t}\mathrm{d}t$ is the Gamma function
$\Gamma(x, y)$	$\int_y^\infty t^{x-1}e^{-t}\mathrm{d}t$ is the upper incomplete Gamma function
$\Upsilon(x, y)$	$\int_0^y t^{x-1}e^{-t}\mathrm{d}t$ is the lower incomplete Gamma function
${}_1F_1(a, b; x)$	$\sum_{r=0}^\infty \frac{(a)_r}{(b)_r}\frac{x^r}{r!}$ is the confluent hypergeometric function of first kind
${}_2F_1(a, b; c; x)$	$\sum_{r=0}^\infty \frac{(a)_r(b)_r}{(c)_r}\frac{x^r}{r!}$ is the Gauss hypergeometric function
$\mathcal{W}_{\lambda,\mu}(x)$	$\frac{x^{\mu+\frac{1}{2}}e^{-\frac{x}{2}}}{\Gamma(\mu-\lambda+\frac{1}{2})}\int_0^\infty \mathrm{e}^{-xt}t^{\mu-\lambda-\frac{1}{2}}(1+t)^{\mu-\lambda+\frac{1}{2}}\mathrm{d}t$ denotes the Whittaker function
$I_\upsilon(x)$	$\sum_{r=0}^\infty \frac{(\frac{1}{2})^{\upsilon+2r}}{r!\Gamma(\upsilon+r+1)}x^{\upsilon+2r}$ denotes the υth-order modified Bessel function of the first kind
$\mathcal{K}_\upsilon(x)$	$\frac{1}{2}\left(\frac{x}{2}\right)^\upsilon\int_0^\infty \frac{e^{-t-\frac{x^2}{4t}}}{t^{\upsilon+1}}\mathrm{d}t$ represents the modified Bessel function of the second kind of order υ
$D_n(x)$	$2^{\frac{1}{4}+\frac{n}{2}}\mathcal{W}_{\frac{1}{4}+\frac{n}{2},-\frac{1}{4}}\left(\frac{x^2}{2}\right)x^{-\frac{1}{2}}$ represents the Parabolic cylinder function
$(x)_k$	$\frac{\Gamma(x+k)}{\Gamma(x)}$ denotes the Pochhammer symbol
$Q(x)$	$\frac{1}{\sqrt{2\pi}}\int_x^\infty \mathrm{e}^{-\frac{t^2}{2}}\mathrm{d}t$ is the Gaussian Q-function

Probability and Statistics

Let X be a random variable (RV).

$\mathbb{E}[\cdot]$	Statistical expectation operator
$\mathcal{P}(\cdot)$	Statistical probability operator
$f_X(\cdot)$	Probability density function (PDF) of a RV X
$F_X(\cdot)$	Cumulative distribution function (CDF) of a RV X
$\mathcal{CN}(\mu, \sigma^2)$	Complex normal distribution with mean μ and variance σ^2

List of Figures

List of Tables

Chapter 1
Introduction

1.1 Overview

In wireless communication, ultra-high data rates, energy efficiency, ultra-reliable and low-latency communication (URLLC), enhanced mobile broadband (eMBB), global coverage and connectivity, and massive machine-type communication (mMTC) are stringent requirements for beyond fifth generation (5G) networks [1]. Figure 1.1 illustrates the interplay between these critical use cases in beyond 5G networks. At the physical layer, the speed of the cellular links has increased manyfold from 50 kbps in 2G systems, 144 kbps in 2.5G systems, approximately 2 Mbps in 3G systems to around 100 Mbps in 4G systems (3GPP-LTE and WiMAX), around 1 Gbps in 5G and around 1 Tbps in sixth generation (6G) systems. Similarly, the speed of indoor wireless local area networks (LANs) has increased from 11 Mbps in IEEE 802.11b to 300 Mbps in IEEE 802.11n within the last 20 years [2]. Although 4G systems provide many services with high data rates, there is still a gap between customers' requirements and the services provided by 4G systems. The goal of 5G networks is to enable seamless global connectivity, allowing communication among people and devices anytime and anywhere. Realizing this vision demands groundbreaking research and novel strategies to overcome traditional limitations, particularly those related to spectrum scarcity. In this regard, the research has shifted to develop advanced methods and groundbreaking wireless technologies to manage massively connected devices. So far, some potential candidates have been proposed to address challenges of 5G, such as, massive multiple input multiple output, millimeter-wave (mm-Wave) communications, ultra dense network, cognitive communication, cooperative communication, non-orthogonal multiple access (NOMA), etc. NOMA has gained attention for its potential to enhance system throughput and support large-scale connectivity. It has been investigated within the 3GPP framework from Release 14 (2015) to Release 16 (2019), with its official adoption for downlink in Release 15, known as evolved universal terrestrial radio access [3–5]. As 5G networks continue to evolve, there is growing momentum toward the comprehensive integration of

V. Bhatia et al., *Advanced NOMA Techniques for Heterogeneous Cellular Networks*, Wireless Networks, https://doi.org/10.1007/978-3-032-09551-0_1

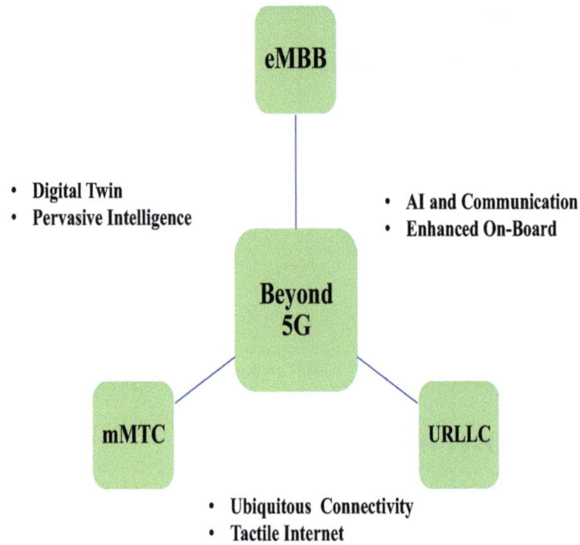

Fig. 1.1 Interplay between beyond 5G use cases

NOMA into future beyond-5G systems. NOMA breaks the orthogonality rule for the multiple access and is able to provide service to multiple users simultaneously on the same resource block (i.e., same frequency, and the same time). Such an multiple access scheme is capable of addressing both, the spectrum scarcity and the need for increased simultaneous connected devices.

1.2 Non-orthogonal Multiple Access

A multiple access (MA) technique allows multiple users to share the same communication channel by employing strategies to mitigate or eliminate mutual interference–typically through separation in frequency, time, or space. These techniques form a critical part of the physical layer and have evolved alongside successive generations of wireless technologies. Each generation introduced more advanced MA methods. For example, the first generation (1G) utilized analog frequency division multiple access (FDMA). With the advent of the second generation (2G), digital techniques became prevalent, and time division multiple access (TDMA) was introduced for efficient time-based multiplexing. Subsequently, code division multiple access (CDMA), proposed by Qualcomm [6], became a key technology in the third generation (3G) networks. However, due to CDMA's limitations in supporting higher data rates, fourth-generation (4G) systems adopted orthogonal frequency division multiple access (OFDMA) [7], a more advanced and efficient scheme. Orthogonal multiple access (OMA) techniques, including OFDMA, remain widely used in 4G systems [8].

NOMA has emerged as a promising alternative to enhance spectral efficiency and enable large-scale connectivity. Recently, NOMA has been proposed for integration into the 3GPP Long Term Evolution (LTE) standard [9, 10]. In LTE-Advanced (LTE-A), NOMA has been studied primarily for two-user scenarios [4]. Generally, NOMA can be divided into two categories: power-domain NOMA (PD-NOMA) and code-domain NOMA (CD-NOMA), the latter often referred to as sparse code multiple access (SCMA). CD-NOMA includes MA solutions relying on low-density spreading [11], SCMA [12], multi-user shared access [13], successive interference cancellation amenable multiple access [14], etc. PD-NOMA and CD-NOMA differ in their basic principle of application. As implied by their names, PD-NOMA utilizes the power domain, whereas CD-NOMA leverages the code domain to achieve non-orthogonal resource allocation. However, PD-NOMA has gained more popularity [15], since no big changes are required at the transmitter as compared to the current 4G technology. This work focuses on the application and analysis of PD-NOMA. PD-NOMA exploits a new dimension of power domain. Thus, it can support multiple users at the same time over a single resource block (RB) by employing successive interference cancellation (SIC). Signals intended for users experiencing varying channel conditions are combined and transmitted together, as illustrated in Fig. 1.2b. Application of NOMA considering two user equipment (UE), termed as cell-center user (CCU) and cell-edge user (CEU), is illustrated in Fig. 1.2b. Based on the order of users' channel gain conditions, SIC is applied to decode their signal [16]. Assume that the base station (BS) allocates more power to the CEU, owing to its weak channel with respect to the BS due to larger distance from the BS as compared to CCU. After receiving the superimposed signal, CEU decodes its own signal by treating CCU's signal as interference. Whereas, CCU performs SIC, i.e., it decodes CEU's signal first and cancels it from the received signal. After SIC, CCU decodes its own signal. At its core, power-domain NOMA (hereafter simply referred to as NOMA) differs significantly from traditional OMA methods such as FDMA, TDMA, CDMA, and OFDMA. By leveraging differences in users' channel conditions, NOMA allocates distinct power levels to multiple users sharing the same resource block. This approach enables simultaneous transmission and leads to more efficient spectrum utilization, making it well-suited for future wireless networks. A theoretical comparison between the conventional OMA technique namely OFDMA and NOMA is depicted in [16]. Theoretical analysis indicates that users supported by NOMA experience higher data rates compared to those using OFDMA. This gain is mainly attributed to the shared use of bandwidth in NOMA, as opposed to the orthogonal division of resources among users in OMA. Furthermore, NOMA leverages users' channel conditions dynamically, enabling it to accommodate multiple users with varying quality-of-service needs [17].

NOMA has gained popularity primarily due to its key benefits, including enhanced spectral efficiency, support for large-scale connectivity, and high compatibility with existing network structures. The enhancement in spectral efficiency stems from NOMA's ability to accommodate multiple users within a single resource block, i.e., the same frequency and time, which indicates efficient utilization of the spectrum and also leads to enhanced sum throughput of the system [16]. In addition, NOMA

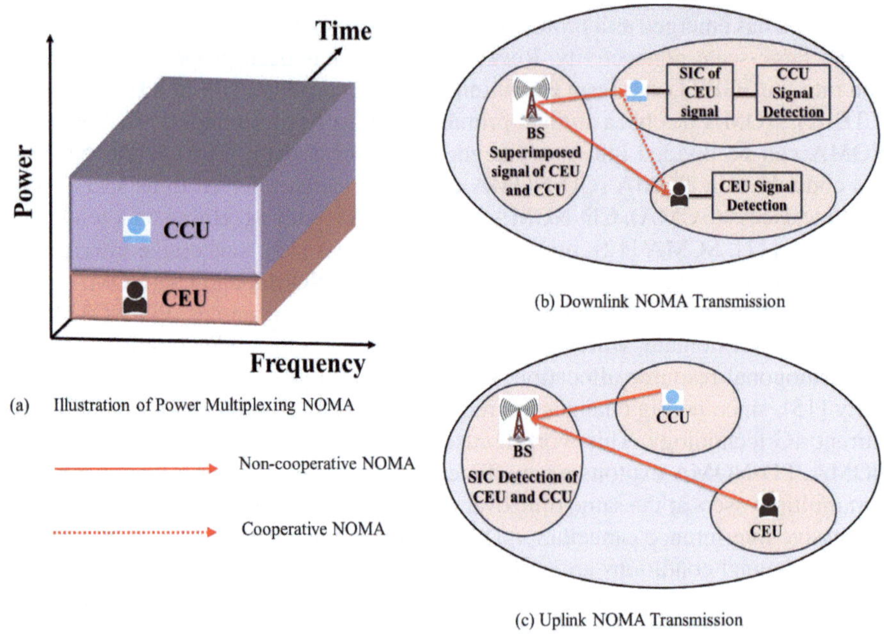

Fig. 1.2 Basic principles of NOMA

enables simultaneous transmission to multiple users, resulting in significantly higher device connectivity. Unlike OMA, which assigns a full resource block to a single user, NOMA allows multiple users to share the same block, thereby increasing user density. Importantly, integrating NOMA into existing 4G systems demands only minor modifications [15]. The theoretical and practical advancements in superposition coding and SIC further ensure NOMA's compatibility and ease of deployment.

1.2.1 Basic Principles

The core idea behind NOMA is to implement multiple access techniques using the power domain (as illustrated in Fig. 1.2a). Unlike the conventional multiple access technologies, NOMA is capable of bringing a new power dimension to perform multiplexing in the existing time/frequency/code domain. In other words, NOMA can be regarded as an add-on technique, which provides the very promising potential to be well integrated with the existing MA paradigms.

1.2.2 Key Technologies of NOMA

NOMA is primarily enabled by two techniques: superposition coding and successive interference cancellation (SIC). These methods are well-established in existing literature and are not entirely novel. As both superposition coding and SIC have advanced in terms of theoretical development and practical implementation, NOMA can now be integrated into next-generation wireless networks with minimal concern for deployment challenges. In this approach, the base station transmits a composite signal formed by superimposing the messages intended for different users. Typically, users are ordered based on their channel gains, either in ascending or descending order. Unlike OMA, which employs power allocation strategies like the water-filling algorithm, NOMA assigns more transmission power to users experiencing weaker channel conditions. This ensures that users with poorer channels can directly decode their messages by treating others as interference. Conversely, users with stronger channels utilize SIC to remove interference from users with weaker channels before decoding their own signals.

Superposition Coding

Originally introduced by Cover in 1972 [18], superposition coding is considered a foundational technique for constructing coding schemes that approach the capacity of scalar Gaussian broadcast channels [19]. It has been theoretically proven that this method can achieve the capacity limits of both Gaussian broadcast channels [20] and more generally, degraded broadcast channels. The central idea behind superposition coding is to encode the data intended for the user with a weaker channel at a lower rate, and then layer the signal for the user with a stronger channel on top of it. Due to the solid foundation laid on the information perspective, researchers begin to apply superposition coding technologies to enormous channels, such as interference channels [21], relay channels [22], multiple access channels [23], and wiretap channels [24]. While the aforementioned contributions sufficiently motivate the use of superposition coding from theoretic perspective, another breakthrough which have been made on superposition coding is to bring this technique from theory to practice [25, 26]. Specifically, [25] designed an experimental platform using a software-radio system to investigate the performance of superposition coding. The set of achievable rate pairs under a packet-error constraint was determined.

Successive Interference Cancelation

SIC is recognized as a powerful technique for efficiently mitigating interference and enhancing network capacity in wireless communications. The SIC process involves the user with a better channel condition first decoding the signal intended for the user with a weaker link. After decoding, the stronger user reconstructs and subtracts the weaker user's signal, thereby removing the interference before decoding its own data. This method allows for improved signal recovery and enables operation close to the Shannon capacity limits for both broadcast and multiple access channels. One of SIC's notable strengths is its relatively low implementation complexity on the receiver side [27]. Due to these advantages, SIC has been extensively explored and applied in various real-world systems such as CDMA [28] and V-BLAST [29]. Beyond these, SIC has also been integrated into diverse scenarios including multi-user MIMO systems [30], multi-hop networks [31], random access protocols [32], and large-scale networks modeled using stochastic geometry [33]. Notably, it has also been adopted in commercial standards like IEEE 802.15.4.

1.2.3 Mathematical Description of NOMA

To better understand the distinction between NOMA and OMA from an information-theoretic standpoint, their performance is analytically compared under a high signal-to-noise ratio (SNR) scenario. Consider a two-user downlink NOMA system, where h_1 and h_2 represent the channel coefficients for User 1 and User 2, respectively, and ρ denotes the transmit SNR at the base station. Without loss of generality, assume $|h_1|^2 \leq |h_2|^2$. According to Shannon's capacity theorem, the achievable rates under OMA are $\frac{1}{2}\log_2(1+\rho|h_1|^2)$ and $\frac{1}{2}\log_2(1+\rho|h_2|^2)$. In contrast, the corresponding rates for NOMA are given by $\log_2\left(1+\frac{\rho\alpha_1|h_1|^2}{1+\rho\alpha_2|h_1|^2}\right)$ for User 1, and $\log_2(1+\rho\alpha_2|h_2|^2)$ for User 2, where α_1 and α_2 are the power allocation factors satisfying $\alpha_1 \geq \alpha_2$ and $\alpha_1+\alpha_2=1$. At high SNR approximation, the sum throughput of OMA and NOMA are $\frac{1}{2}\log_2\left(\rho|h_1|^2\right)+\frac{1}{2}\log_2\left(\rho|h_2|^2\right)=\log_2\left(\rho\sqrt{|h_1|^2|h_2|^2}\right)$, and $\log_2\left(\rho|h_2|^2\right)$, respectively.It has been observed that in the high SNR regime, NOMA achieves a notably higher sum rate compared to OMA, particularly when there is a significant disparity in users' channel conditions. From a mathematical viewpoint, this performance gain arises because dividing resources such as time or frequency (e.g., by a factor of 1/2) outside the logarithmic function results in greater loss than allocating power within the logarithm.

1.2.4 Downlink and Uplink NOMA Transmission

In downlink NOMA transmission, the base station utilizes superposition coding to transmit a combination of user signals, while users employ SIC to mitigate interference. Specifically, as depicted in Fig. 1.2b, the cell-center user (CCU) applies SIC to eliminate interference from the composite signal, whereas the cell-edge user (CEU) decodes its message by treating the signal intended for the CCU as interference. This work concentrates on downlink NOMA scenarios, primarily because its practical variant, known as multi-user superposition transmission (MUST), has already been adopted in LTE-A standards, making it a strong candidate for 5G systems.

In contrast to downlink NOMA, uplink NOMA requires the base station to initially send control signals to multiple users for appropriate power allocation. Following this, users simultaneously transmit their individual data to the base station using the same orthogonal resource block, as illustrated in Fig. 1.2c.

1.2.5 Cooperative NOMA

The fundamental principle of cooperative NOMA lies in utilizing strong NOMA users as decode-and-forward (DF) relays to assist weaker users. Consider, for instance, the two-user downlink scenario illustrated in Fig. 1.2b. Cooperative NOMA involves a two-phase transmission process. In the initial phase–known as the direct transmission phase–the data is transmitted similarly to standard non-cooperative NOMA (represented by the solid line in Fig. 1.2(b)). In the second phase–referred to as the cooperative phase–the strong user (CCU), having already decoded the weak user's message using successive interference cancellation (SIC), forwards this message to the weak user (CEU) by employing a DF relaying protocol (indicated by the dashed line in Fig. 1.2b). This innovative approach, as introduced in [10], has garnered considerable research interest due to its potential to combine the strengths of SIC and cooperative relaying.

Compared to traditional NOMA systems, cooperative NOMA offers several notable benefits: (1) Since the strong user in NOMA already decodes the weak user's message through SIC, it is intuitive and efficient to incorporate the DF relaying mechanism. (2) The reliability of the weak user is substantially improved, promoting a more balanced performance across users. (3) Cooperative NOMA provides better resistance to multipath fading by offering higher diversity gain for the weak user, thus improving overall link robustness.

1.3 Heterogeneous Networks

As cellular networks face growing data traffic and current link efficiency nears its theoretical limits, enhancing spectral efficiency has become increasingly important. One potential solution is to deploy additional macro base stations (MBSs). However, introducing more MBSs into an already densely populated network can lead to elevated inter-cell interference [34]. Moreover, the installation of extra MBSs in urban or congested areas can incur substantial costs, particularly due to the challenges of acquiring new deployment sites. An alternate strategy to deal with the ongoing issue is deploying smaller base stations, i.e., BS with lower power (250 mW to approximately 2 W) and smaller transmission range, which also poses lower deployment cost. Such small base stations include pico base stations, femto base station (FBS), etc. A HetNet, widely adopted in LTE-A systems, can be formed by overlaying small base stations onto the MBS infrastructure. In such a setup, MBSs operate at high transmit power levels (up to 40 W), while femto base stations (FBSs) transmit at significantly lower power and are often deployed in an ad hoc manner. The inclusion of small cells enhances network coverage and boosts capacity through increased spatial reuse. Unlike pico base stations, which are primarily intended for outdoor environments, FBSs are typically used indoors. Since FBSs are generally installed by end users, they can configure their operational mode, deciding which users are permitted to access them. These access modes include open access and closed access [35, 36]. In closed access mode, only a predefined group of users can connect to the FBS. In contrast, open access mode allows any user in the network to connect to the FBS without restriction.

1.3.1 Stochastic Geometry for Heterogeneous Networks

Stochastic geometry provides a mathematical framework for analyzing random spatial patterns and topologies [37–40]. Within the context of wireless networks, this approach involves assigning probability distributions to the spatial positions of transmitters and receivers. In essence, the locations of these nodes are modeled as realizations of point processes–random processes defined over Euclidean space (e.g., $\mathbb{R}^2$). For many years, point process models have served as a foundational tool in the analysis of wireless ad hoc networks, offering insights into network performance [41–44]. In the realm of cellular networks, early works such as [45, 46] employed the Poisson point process (PPP) to represent the spatial distribution of base stations for modeling both traffic dynamics and economic factors. Building on this foundation, [47] analyzed key performance metrics–such as coverage probability and data rate–in single-tier cellular networks using PPP-based BS distributions. Furthermore, [35] extended this framework by proposing an analytical model based on PPP for heterogeneous cellular networks comprising both macro base stations and small cell base stations (SBSs).

In a stochastic geometry-based cellular network model, the BSs and user locations are treated as realizations of some random point pattern, also known as point processes. Then the performance metrics of the network (such as the coverage and rate of an arbitrary user) are evaluated as functionals of these point processes. Stochastic geometry plays a crucial role in simplifying point process functionals–such as sums, products, or combinations thereof–into tractable forms, which often result in integrals over $\mathbb{R}^2$. A widely adopted modeling approach involves using a homogeneous Poisson Point Process (PPP) to represent the spatial distribution of base stations (BSs) and users. This effectively assumes that both users and BSs are randomly and independently scattered across the plane with uniform density. The use of PPP-based models has become increasingly popular in the study of cellular networks due to their analytical tractability and ability to yield insightful performance metrics.

Tractability: This modeling approach allows for the derivation of analytically tractable expressions–often in closed-form–for key performance indicators such as coverage probability and data rate [47]. Furthermore, the framework can be extended to model multi-tier heterogeneous networks (HetNets), incorporating various types of base stations, including macro, pico, and femto cells [35]. Consequently, this enables the evaluation of a broad range of network deployment scenarios within a unified analytical framework [48, 49].

Flexibility: These models exhibit high versatility, making them suitable for evaluating the performance of various advanced wireless technologies, including energy-harvesting base stations [50], MIMO-enabled systems [51], localization methods [52], and millimeter-wave (mmWave) communication [53], among others. The development of PPP-based modeling approaches for cellular networks has reached a considerable level of maturity. For comprehensive tutorials and detailed surveys of earlier contributions, readers are encouraged to consult [48, 54, 55]. Over the past ten years, Poisson Point Processes have served as a fundamental tool in stochastic geometry-based models of cellular networks.

1.3.2 NOMA Enabled Heterogeneous Networks

NOMA and heterogeneous networks (HetNets) have each gained significant attention in wireless communications. When integrated, these technologies offer enhanced performance benefits, as outlined below [56].

- Improved transmit SINR: Although HetNets provide several advantages, the dense deployment of base stations often leads to increased interference, which can degrade the SINR experienced by users. Incorporating NOMA into these networks allows for more effective management of various types of interference–such as inter-tier, intra-tier, and intra-group interference–thereby enhancing the overall SINR at the user end.
- Improved fairness: Fairness among the users is very important for the HetNets, especially while investigating efficient resource allocation in multi-tier networks. NOMA provides an efficient way to address fairness by allocating more power to the user with poorer channel condition.

MBS typically cater to users with similar service demands, often allocating power evenly across all users. In contrast, femto base stations (FBS) function more like wireless LAN routers and are designed to support a high density of users with diverse requirements–ranging from low-data-rate IoT devices to bandwidth-intensive applications like HD video streaming [57]. These varied demands necessitate adaptive power allocation. NOMA facilitates this by allowing dynamic power sharing among users with different data needs. Consequently, implementing NOMA in the FBS layer is generally more suitable than in the MBS layer, particularly in densely populated network environments. Furthermore, NOMA has a significant role in HetNets in situation where users are offloaded to the open access FBS tier, and the FBS happens to be fully-loaded [58] with its registered subscribers. Since a fully-loaded FBS cannot allocate orthogonal resource to the incoming offloaded user, NOMA plays a vital role in serving the offloaded user by pairing it with the appropriate pairing user.

1.4 Point Process

Traditional cellular networks often rely on a grid-based layout for their design, where base stations–especially MBSs–are positioned according to a fixed grid structure. While widely adopted, these models are highly idealized and lack analytical tractability. As a result, evaluating performance metrics like outage probability and data rate typically requires extensive system-level simulations. Additionally, modeling interference remains a significant challenge, even in single-tier networks. In particular, analyzing interference under the conventional hexagonal grid framework either necessitates computationally heavy Monte Carlo simulations [6] or leads to inaccuracies due to oversimplified assumptions [59]. Thus, a more realistic model with tractable analysis is required, especially for modeling the multi-tier BSs of the HetNets. In real-world deployments, base station (BS) placements rarely follow a rigid grid due to varying capacity requirements across different regions such as residential neighborhoods, parks, suburban, and rural areas. Observations of BS layouts from various locations reveal that their distribution appears largely random, deviating significantly from the idealized grid-based models. This spatial irregularity is observed in both single-tier and multi-tier cellular networks, making traditional models unrealistic for modern deployments. Efficient design of heterogeneous networks (HetNets) demands careful interference management, highlighting the need for accurate and analytically manageable interference models. Stochastic geometry emerges as a powerful framework to address this challenge. It not only captures the spatial randomness of BS locations but also facilitates tractable performance analysis [54, 60, 61]. Within this framework, various spatial point processes are used to represent BS or user locations, with each process tailored to reflect specific real-world configurations. A point process refers to a stochastic model that describes the random placement of points (such as BSs or users) in a spatial domain. Comprehensive reviews of different point processes can be found in [40, 54, 61]. Notable point processes utilized in wireless network analysis include the Poisson Point Process (PPP) [35, 58, 62], the Hard-Core Point Process [63–65], and the Poisson Cluster Process [66–68]. A brief introduction to the common point processes is given below:

- Poisson point process: A point process $\Omega = \{x_i, i = 1, 2, 3, \ldots\} \subset \mathbb{R}^d$ is called a PPP if and only if the number of points inside any compact set $B \subset \mathbb{R}^d$ is a Poisson random variable, and the numbers of points in disjoint sets are independent, where d represents number of dimensions.
- Hard core point process (HCPP): An HCPP is a repulsive point process where the distance between two co-existing points is always greater than a predefined hard core parameter r. A point process $\Omega = \{x_i, i = 1, 2, 3, \ldots\} \subset \mathbb{R}^d$ is an HCPP if and only if $\| x_i - x_j \| \geq r \ \forall x_i, x_j \in \omega, i \neq j$, where $r > 0$ is a predefined hard core parameter.
- Poisson cluster process (PCP): The PCP is used to model random patterns produced by random clusters. The PCP is constructed from a parent PPP $\Omega = \{x_i, i = 1, 2, 3, \ldots\} \subset \mathbb{R}^d$ by replacing each point $x_i \in \Omega$ with a cluster of points

$C_i, \forall x_i \in \omega$, where the points in C_i are independently and identically distributed in the spatial domain.

For more formal and detailed definitions of the point processes, readers can refer to [54, 61].

1.5 Simultaneous Wireless Information and Power Transfer

In wireless communication networks, energy harvesting (EH) is a new paradigm that enables terminals to recharge their batteries using the received radio-frequency (RF) signal [69]. The energy constraint issue in wireless networks can be resolved by RF-EH. EH is the process of transforming RF energy into electrical energy. In contrast to traditional battery or grid-powered communications, the EH offers several unique advantages and promising benefits for future wireless communications, such as self-sustaining functionality, reduction of carbon footprint, wireless nodes that don't require battery replacement, and many more [70]. As a result, EH in wireless networks is becoming increasingly popular in a variety of applications, such as medicinal implants, remote environmental monitoring, etc. Particularly in the RF-EH, simultaneous wireless information and power transfer (SWIPT) provides simultaneous transfer of information and power. SWIPT-enabled receiver architectures such as time switching (TS) and power splitting (PS) are commonly used.

1.5.1 Time Switching

TS architecture, also known as co-located receiver architecture, shares the same antenna for EH and information reception [71]. As illustrated in Fig. 1.3, the receiver employed in this architecture includes an RF energy harvester, an information decoder, and a switch that alters the system's receiving antenna. Based on a TS sequence, the receiver antenna switches between the energy harvester and information decoder circuit periodically. Further, the TS receiver also requires accurate information/energy scheduling and time synchronization. The TS factor can be optimized to achieve optimum performance.

1.5.2 Power Splitting

The PS receiver divides the received signal into two power streams of different power levels with a certain PS ratio before signal processing is performed at the receiver [71]. To enable simultaneous EH and information decoding, both power streams are then delivered to an information decoder and energy harvester as depicted in Fig. 1.4.

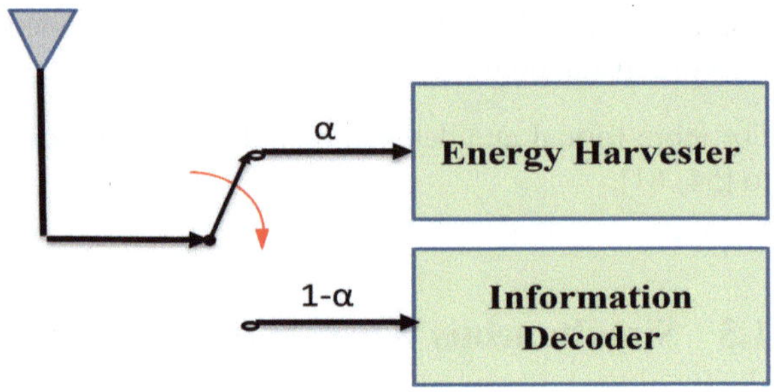

Fig. 1.3 Time switching receiver architecture. α is the time switching coefficient which describes the portion of the time to be dedicated for energy harvester and the information decoder

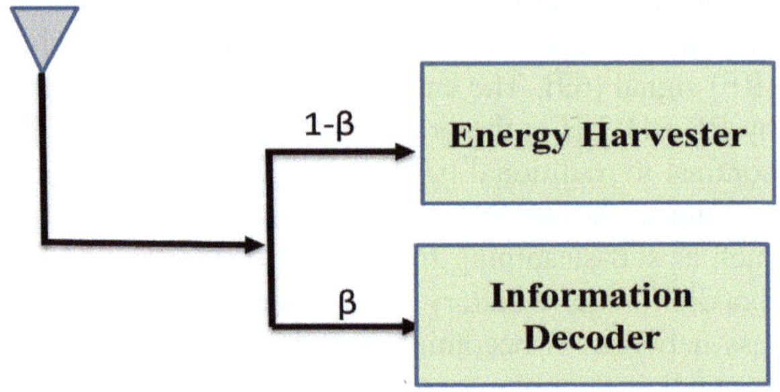

Fig. 1.4 Power splitting receiver architecture. β is the power splitting coefficient, such that β times the received power is used for information decoding and the rest i.e., $1 - \beta$ times received power is used for EH

Further, by varying PS ratios, the information rate and the harvested energy can be balanced according to the system requirements. Furthermore, the overall performance can also be improved by optimizing the PS ratio.

1.6 Reconfigurable Intelligent Surface

Reconfigurable intelligent surface (RIS) has emerged as a transformative technology for next-generation wireless communication systems owing to its ability to intelligently reconfigure the wireless propagation environment. Structurally, an RIS is a planar metasurface composed of a large number of reflecting elements (REs), each of which can independently manipulate the phase and/or amplitude of the incident electromagnetic wave under the control of a central RIS controller. This reconfigurability enables precise reflective beamforming, thereby enhancing signal strength, coverage, and interference management in a cost- and energy-efficient manner.

A conventional passive RIS is comprised of numerous passive REs that are capable of imposing tunable phase shifts on the incident signal. While passive RIS solutions are attractive due to their low hardware complexity and energy consumption, they inherently suffer from the multiplicative fading effect, which stems from the cascaded path losses of the transmitter–RIS and RIS–receiver links. This limitation reduces the overall effectiveness of passive RIS in scenarios requiring long-range coverage or strong signal strength.

To mitigate this drawback, active RIS architectures have been proposed, wherein each RE is equipped with active reflection-type amplifiers, such as asymmetric current mirrors or current-inverting converters. Unlike passive RIS, active RIS not only manipulates the phase of the incident signal but also amplifies it, thereby compensating for the severe path loss inherent in cascaded channels. This amplification significantly boosts signal power, extending coverage and improving spectral efficiency, provided that the additional thermal noise introduced by active elements remains within tolerable limits.

The integration of RIS with NOMA offers substantial performance gains by dynamically shaping the wireless propagation environment in favor of users. In particular, active RIS-assisted NOMA systems can ensure robust connectivity for cell-edge and mobile users who often experience degraded performance due to obstacles or unfavorable channel conditions. By adaptively controlling the reflective beam patterns, RIS enhances power-domain user separation in NOMA systems, thereby improving spectral efficiency, fairness, and user connectivity.

1.7 Wireless Communication Channel Model

The radio-wave propagation through the wireless channel can be primarily characterized by two phenomena, multi-path propagation, and shadowing. Mathematically,finding the exact analysis of the fading phenomenon is a complex task. However,through researchers' constant efforts, statistical models have been developed to characterize the channel under fading. The radio environment between the transmitter and receiver tends to fluctuate during transmission of the signal. Based on the fluctuations, there exist various fading models that statistically describe the nature of the multi-path propagation. Few of the commonly used fading models include Rayleigh fading, Nakagami-m fading, Ricean fading, and Weibull fading [72, 73]. This thesis utilizes Rayleigh distribution for the small scale fading as discussed below:

- **Rayleigh distribution**: The Rayleigh distribution is the most popular method for describing the radio channel's statistical behavior. It is operated in the case where multipath propagation exists without a dominant line-of-sight (LoS) path between the end users. Due to the constructive or destructive interference of the multipath components, the in-phase and quadrature-phase components of the received signal are modeled with a zero mean complex Gaussian random process. Thus, the amplitude of the received signal is Rayleigh distributed. The PDF of the Rayleigh distributed signal can be expressed as

$$f_{\gamma}(x) = \frac{1}{\Omega} \exp\left(-\frac{x}{\Omega}\right), \quad x \geq 0 \tag{1.1}$$

where Ω represents the average SNR.

1.8 Performance Metrics

To examine the performance of SWIPT-enabled cooperative NOMA based HetNets, several performance metrics are used for different modulation schemes. To fix various design issues of wireless communication systems, these performance measures are used. Commonly used performance metrics are instantaneous SNR, outage probability, system throughput, energy efficiency, ergodic capacity, and average symbol error rate (ASER) [73].

1. **Instantaneous SNR**: The instantaneous SNR is a basic performance metric that is used to quantify the signal corruption due to noise. Instantaneous SNR is related to data detection as it is measured at the output of the receiver and is an excellent indicator of the overall fidelity of the communication system. Instantaneous SNR can be expressed as

$$\gamma = \frac{P|h|^2}{\sigma^2}, \tag{1.2}$$

where P, h, and σ^2 represent the transmit power, channel parameter, and noise variance, respectively. Due to the multipath fading in wireless communication, average SNR is a more appropriate performance metric than instantaneous SNR. Average SNR is the statistical averaging over the probability distribution of the fading and is given as $\Omega = \mathbb{E}[\gamma]$, where $\mathbb{E}[\cdot]$ represents the statistical expectation operator.
2. **Outage Probability**: Outage probability is one of the important performance metric that depicts link failure probability and is mainly used in the case of a slow-fading scenario. It is defined as the probability that the received end-to-end instantaneous SNR (γ) of the considered system lies below a predefined threshold (γ_{th}), which can be given as

$$\begin{aligned}\mathcal{P}_{\text{out}}(\gamma_{\text{th}}) &= \Pr\left[\gamma < \gamma_{\text{th}}\right] \\ &= \int_0^{\gamma_{\text{th}}} f_\gamma(x)\mathrm{d}x,\end{aligned} \tag{1.3}$$

where $f_\gamma(\cdot)$ represents the PDF of γ, $\gamma_{\text{th}} = 2^{r_{\text{th}}} - 1$, and r_{th} represents the threshold data rate.
3. **System throughput**: The system throughput is one of the important performance metrics to characterize spectrum utilization. It can also be referred to as mean spectral efficiency. Based on the derived expression of outage probability, the system throughput can be formulated as

$$\tau = \left[1 - \mathcal{P}_{\text{out}}(\gamma_{\text{th}})\right] r_{\text{th}}. \tag{1.4}$$

4. **Energy efficiency**: Energy efficiency is an important performance metric in the wireless communication system to achieve the vision of green communication

systems. It is defined as the overall data transferred to the overall consumed energy. The overall data transferred is referred to as system throughput. Energy efficiency can be expressed as

$$\eta_{\text{EE}} = \frac{\mathcal{T}}{P}. \tag{1.5}$$

5. **Ergodic capacity**: Ergodic capacity quantifies the ultimate reliable communication limit over fading channels. Instantaneous capacity (measured in bps/Hz) is defined as the maximum rate achieved by the communication channel and can be determined as $C = \log_2(1 + \gamma)$. Hence, ergodic capacity is obtained by averaging the instantaneous capacity over the PDF of the instantaneous SNR (γ), which is defined as

$$\begin{aligned} C_{\text{e}} =& \mathbb{E}\left[\log_2(1+\gamma)\right] \\ =& \int_0^{\infty} \log_2(1+x) f_{\gamma}(x)\text{d}x. \end{aligned} \tag{1.6}$$

6. **ASER**: ASER is an important performance metric for wireless communication systems that can be determined by averaging the symbols with error at the receiver. It can be given as

$$\mathcal{P}_e = -\int_0^{\infty} \mathcal{P}_s^{'}(e|x)\mathcal{P}_{out}(x)\text{d}x, \tag{1.7}$$

where $\mathcal{P}_s^{'}(e|x)$ represents the first order derivative of the conditional symbol error probability ($\mathcal{P}_s(e|x)$) for the received SNR.

1.9 Motivation

As a promising candidate for future 6G systems, NOMA transmission offers a multitude of advantages. One of its standout features is its high spectrum efficiency, a vital performance metric in wireless networks. NOMA achieves exceptional spectrum efficiency by allowing multiple users to share a single resource block, be it time, frequency, or code. Unlike traditional power allocation policies, NOMA allocates more power to weaker users, striking a crucial balance between fairness and system throughput. Moreover, NOMA's ultra-high connectivity capabilities align perfectly with the 6G vision of connecting billions of smart devices, such as those in the IoT. By capitalizing on its non-orthogonal characteristic, NOMA efficiently supports numerous devices with fewer RBs compared to conventional OMA methods. In addition, NOMA exhibits excellent compatibility as it can complement existing OMA techniques like TDMA, FDMA, CDMA, or OFDMA by introducing a new power dimension. With the maturation of superposition coding and SIC technologies in both theory and practice, NOMA holds the promise of seamless integration into existing wireless systems. Furthermore NOMA enabled small cells is also a potential enabler for 6G systems.

Motivated by the flexibility and compatibility of NOMA, this thesis aims at investigating the application of NOMA with recent advanced technologies, e.g., HetNets, and SWIPT. Various stochastic tools are utilized to evaluate the performance of the proposed system models. The application of SWIPT in networks employing cooperative NOMA is investigated in a multi-tier downlink scenario with MBSs and FBSs. The NOMA is applied at the FBS tier and performance analysis is performed at the femto users distributed uniformly in the coverage area of FBS. The near user employs SWIPT using superimposed NOMA signal and the interference from the adjacent base stations. The harvested energy is utilized for cooperative transmission to the far user. Besides, the outage probability and system throughput is evaluated at the user pair served by the FBS using NOMA. Moreover, for interference management at the small base station tier, medium access channel (MAC) protocol is applied by utilizing carrier sensing. The performance of the resulting network is investigated using repulsive point process. Moreover, since the TS architecture require less hardware complexity compared to the PS architecture. This motivates to explore a NOMA enabled HetNet with a TS protocol at the cooperating user for EH. Furthermore, the application of NOMA involves pairing of users with varying channel conditions, consideration of user fairness is also crucial. This motivates to ensure the users' fairness, and evaluate the proportional fairness index. Moreover, the rectennas in the energy harvester include non-linear components (e.g., diodes and capacitors), hence the input/output relationship is non-linear. Therefore, this thesis studies the performance of EH assisted cooperative NOMA in a two-tier HetNet with a realistic non-linear EH model. Furthermore, the SWIPT based systems are appropriate for the short distance communication. To overcome this limitation, PB is used to supply power for wireless devices. Hence, this thesis studies the performance of power beacon (PB) assisted downlink cooperative NOMA in a wireless powered communication network (WPCN) with realistic non-linear EH model.

1.9.1 Related Works

The concept and practical consideration on NOMA is investigated in [16]. Performance of downlink NOMA with randomly deployed users is studied in [74], which proves the superior ergodic rate achieved by NOMA over OMA. Sub-channel assignment, power allocation, and user scheduling in NOMA is studied in [75]. In [17], impact of user pairing in NOMA is studied and [76] studies a low complexity user pairing algorithm. In [77], cooperation techniques are studied from NOMA perspective. Authors in [78] analyze NOMA in a cognitive underlay network wherein the interference from primary network to secondary users is considered. A joint fractional time allocation and beamforming in NOMA network is investigated in [79]. A fair power allocation scheme for NOMA is discussed in [80]. The authors in [17] prove that in a NOMA group, a user with better channel condition, i.e., a strong user achieves higher throughput in comparison to a user with poorer channel condition, i.e., a weak user. In [81], a coordinated superposition coding scheme is presented, which can improve the CEU's rate, without sacrificing the rate of the users, which are close to the BS. In this direction, cell boundary user's performance is also analyzed in [82], and user fairness in NOMA is dealt in [83], where a dynamic user clustering problem is formulated from a fairness perspective, and power allocation coefficients for the users in each cluster is optimized. Furthermore, to eliminate unfairness in NOMA networks, cooperative NOMA is used wherein a nearby user (i.e., a strong user) is treated as a relay to assist a distant user (i.e., a weak user) as studied in [77]. The authors in [84] use energy harvesting at the strong user to assist the weak user using cooperative NOMA. As a further advance, the fairness issue of NOMA has been addressed in [85], by examining appropriate power allocation policies among the NOMA users. In [86], the sum rate and outage probability of NOMA with partial CSI is investigated. The work in [87] investigate NOMA using stochastic geometry.

1.9.2 NOMA and Cooperative Transmission

Recent advancements in NOMA with cooperative relaying systems predominantly utilize two relay protocols: amplify-and-forward (AF) and decode-and-forward (DF). For AF-based schemes, the studies in [88, 89] examined the outage behavior of NOMA-enabled downlink systems, where both the base station and users were equipped with either single or multiple antennas. These works also highlighted the performance benefits of NOMA compared to OMA in both scenarios. With respect to DF relaying, in addition to its effective integration with cooperative NOMA as presented in [77], a two-stage DF relaying protocol was introduced in [90]. Beyond enhancing transmission coverage, cooperative communication also plays a vital role in achieving spatial diversity to counter fading effects. Coordinated Multi-Point (CoMP) transmission has been recognized as an efficient method for interference reduction, especially for improving service at the cell edges. The application of NOMA within CoMP frameworks was first explored in [81], where a coordinated superposition coding technique combined with Alamouti coding was proposed to enhance spectral efficiency. Furthermore, [91] introduced a novel opportunistic NOMA strategy for CoMP systems, demonstrating superior performance over traditional joint-transmission NOMA approaches. Kim and Lee [92] considered a coordinated direct and relay transmission scheme, where the BS communicated with a near user and a relay simultaneously with invoking NOMA technique during the first phase, while communicated with a far user with the aid of the relay. Vien et al. [93] proposed a NOMA based power allocation for a downlink cloud radio access network (CRAN) scenario, which can be also regarded as a coordinated transmission scenario where all the BSs are controlled by a cloud.

1.9.3 Wireless Energy Harvesting and NOMA

In recent years, extracting energy from ambient RF signals has gained significant traction as a practical means to power low-energy communication systems autonomously [94]. This trend is largely driven by technological advancements in low-power hardware, seen in both industrial developments [95] and academic research [96]. As a result, RF energy harvesting is emerging as a viable solution for powering future low-energy applications, particularly in networks composed of power-constrained devices like wireless sensor nodes [97, 98]. The concept of SWIPT, first introduced in [99], has revitalized research aimed at developing energy-efficient communication systems. Although the initial concept in [99] assumed the concurrent extraction of energy and information from the same RF signal, practical limitations prompted the proposal of two implementable receiver architectures TS and PS in the context of MIMO systems [100].

While SWIPT systems are well documented in the literature, attention has recently shifted towards cooperative relaying systems employing either AF or DF strategies. For AF relaying, both TS and PS-based relaying protocols were introduced in [101]. In contrast, for DF relaying, an antenna switching-based SWIPT protocol aimed at reducing system complexity was proposed in [102]. Moreover, a novel DF relaying scheme that enables energy harvesting in underlay cognitive radio (CR) networks was proposed in [103], allowing secondary users to harvest energy from primary transmissions. The integration of SWIPT in DF-based cooperative networks with randomly located relays was further analyzed in [104] using stochastic geometry, focusing on a setup with multiple sources and a common destination. Additionally, [77] studied a scenario involving multiple source-destination pairs that communicate through a single energy-harvesting relay. Regarding NOMA systems related with SWIPT, in [105], the uplink NOMA transmission considering energy constrained users which can harvest energy from the BS with adopting "harvest-then-transmit" protocol.

1.10 Book Outline, and Contributions

The flow of the book is organized into 7 chapters, which are briefly described with their contributions as follows:

Chapter 1. Introduction: In this chapter, a brief introduction to the NOMA, cooperative NOMA, various performance metrics, simultaneous wireless information and power transfer, HetNets, and finally, the motivation and major contributions of the work presented in the book are provided.

Chapter 2. Joint Impact of HPA Non-linearity and Imperfect SIC in NOMA Enabled HCN: In heterogeneous cellular networks (HCNs), a multi-tier framework comprising MBS and pico base stations (PBS) is deployed to meet the growing demand for high data rates and better user experience. To accommodate the increasing number of users expected in future sixth-generation (6G) networks, NOMA has emerged as a promising spectral efficiency technique. However, to enhance power efficiency and reduce outage probability, base station power amplifiers are often operated near their saturation limits, leading to non-linear distortions (NLD). Hence, it becomes crucial to evaluate the performance of NOMA-enabled HCNs (HCN-NOMA) in the presence of such distortions caused by high-power amplifiers (HPAs).

This work utilizes stochastic geometry to perform a comprehensive performance analysis of HCN-NOMA systems under the influence of NLD and imperfect successive interference cancellation (i-SIC). A comparative evaluation of OMA and NOMA is conducted under both ideal and imperfect SIC conditions in multi-tier HCNs. Key performance metrics such as outage probability, ergodic capacity, system throughput, and energy efficiency are investigated. Results indicate that NOMA offers up to 96% performance improvement over OMA in the non-linear operating region of HPAs. Moreover, even under i-SIC conditions, NOMA achieves performance gains of up to 93% compared to OMA, highlighting its robustness in non-ideal scenarios.

Chapter 3. Performance of NOMA-empowered HCN: A Stochastic Geometry Approach Considering Imperfect SIC and CSI: This chapter investigates the performance of NOMA in multi-tier heterogeneous cellular networks (HCNs) under practical conditions of imperfect successive interference cancellation (i-SIC) and imperfect channel state information (i-CSI) using stochastic geometry. A comparison is made between OMA and NOMA systems, analyzing key performance metrics such as outage probability, asymptotic outage probability, system throughput, and energy efficiency, considering both ideal and i-SIC scenarios. The chapter presents several key contributions, including a comparison of HCN-NOMA and OMA systems with perfect and i-SIC scenarios, along with the derivation of closed-form outage probabilities and asymptotic outage probabilities for i-CSI systems. Furthermore, the study explores the interplay between i-SIC and i-CSI, providing valuable insights into their collective impact on spectral and energy efficiencies.

Chapter 4. SWIPT Enabled Cooperative NOMA in Heterogeneous Networks: This chapter presents an analytical framework for a heterogeneous downlink network incorporating SWIPT and cooperative NOMA. The proposed model focuses on enhancing the performance of the CEU through assistance from the CCU. To sustain its cooperative function without depleting its battery, the CCU is equipped with SWIPT capabilities. It harvests energy from both the superimposed transmitted signal and the cumulative interference present in the HetNet environment. Closed-form expressions for the outage probabilities of both CCU and CEU are derived. Additionally, the chapter examines analytical formulations for system throughput and user fairness. The impact of imperfect CSI on the overall system performance is also analyzed. A comparative analysis is carried out with conventional energy harvesting models–where energy is extracted solely from the superimposed signal–highlighting the advantages and novel insights offered by the proposed scheme.

Chapter 5. Performance of HCN NOMA Under Im-N: The presence of impulsive electromagnetic interference in smart homes, grids, and industrial IoT applications challenges conventional Gaussian noise-based analyses. Finally, this thesis evaluates the performance of HCN-NOMA under additive white Gaussian noise (AWGN) and impulsive noise (Im-N) using stochastic geometry. A comparative analysis with OMA is conducted, revealing NOMA's superior spectral and energy efficiency, even under Im-N. Mathematical modeling incorporates a PPP for spatial distribution of nodes, with closed-form expressions derived for key performance metrics. These findings offer valuable insights into the interplay between impulse noise occurrence probability and impulsive-to-Gaussian noise power ratio, illustrating their combined impact on system efficiency. The findings emphasize the advantages of NOMA over OMA, identifying specific SNR ranges that maximize performance gains under both AWGN and Im-N conditions.

Chapter 6. **Performance of RIS-Aided NOMA Network**: The study examines the performance of a downlink NOMA network assisted by an active reconfigurable intelligent surface. Analytical expressions for the outage probability (OP) and ergodic capacity are derived, utilizing the Meijer-G function. For the OP, asymptotic analysis is performed to determine the system's diversity order. Additionally, closed-form expressions for the average symbol error rate are derived for advanced modulation schemes, including rectangular, cross, and hexagonal quadrature amplitude modulation (QAM) schemes. The effects of various system parameters, such as the number of reflecting elements, their amplification gain, thermal noise, and other factors, are thoroughly investigated. A performance comparison with OMA is also presented.

Chapter 7. **Future Works**: All the contributions of the thesis have been summarized in this chapter, and important insights and conclusions have been presented. Further, the scope for future works is also discussed.

Chapter 2
NOMA-HCNs: HPA and SIC Imperfections Impact

The current wireless communication paradigm confronts multiple challenges—namely, the demand for higher data rates, broader user coverage, and enhanced energy efficiency, reduced latency, and lower cost per bit transferred [106]. To address these demands, research focus has shifted toward 6G, where ultra-dense networks (UDNs) comprised of micro-, pico-, and femtocells offer a promising solution. Owing to their low cost and power requirements, densely deploying these small cells markedly increases network capacity and supports a larger user population. As user density grows, the finite spectral and power resources render orthogonal allocation impractical. Assigning each user a dedicated frequency band and time slot becomes especially challenging under poor channel conditions or diverse rate requirements. Consequently, power-domain NOMA has emerged as an attractive multiple-access scheme for future UDNs [107, 108], enabling multiple users to share the same time–frequency resource with differentiated power levels. Compared to conventional OMA techniques (e.g., FDMA, TDMA, or CDMA), NOMA delivers superior spectral efficiency and supports a higher number of simultaneous connections. Looking forward, several research directions promise to further advance NOMA's capabilities. Integrating wireless power transfer (WPT) with NOMA can revolutionize energy-efficient network design, while pairing NOMA with cognitive radio networks (CRNs) facilitates dynamic spectrum allocation to satisfy massive connectivity requirements. Extending NOMA to heterogeneous 6G scenarios offers enhanced connectivity and network optimization, and exploring its application in visible light communications (VLC) could redefine post-5G paradigms with unprecedented data rates [106]. Moreover, combining NOMA with multiple-antenna techniques yields significant gains. Whereas massive MIMO-OMA may stumble when user counts exceed antenna numbers, massive MIMO-NOMA can overcome this bottleneck, meeting the stringent connectivity and spectral-efficiency demands of future networks [15]. When user power coefficients and rate allocations are optimally configured, NOMA outperforms OMA in both ergodic sum rate and outage probability [109, 110]. At the receiver, successive interference cancellation (SIC) enhances quality of service (QoS) by decoding weaker users' signals first, thereby lowering error rates and elevating

V. Bhatia et al., *Advanced NOMA Techniques for Heterogeneous Cellular Networks*, Wireless Networks, https://doi.org/10.1007/978-3-032-09551-0_2

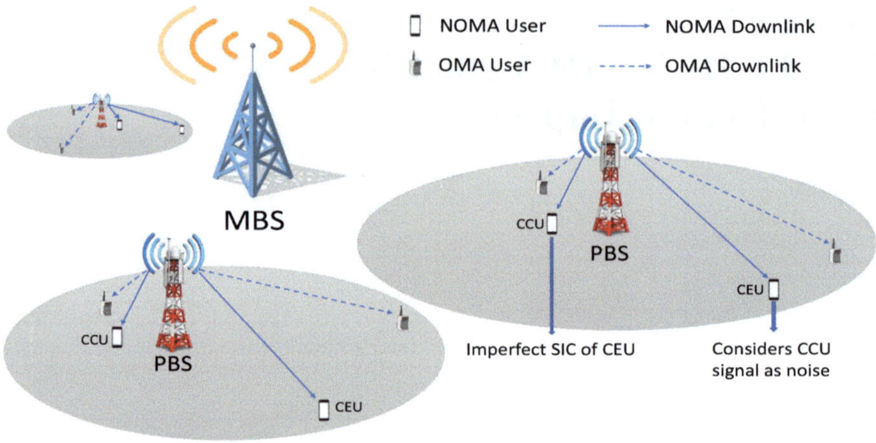

Fig. 2.1 System model for HCN-NOMA

data rates for users in poor channel conditions [111]. NOMA's versatility also permits integration with emerging technologies such as intelligent reflecting surfaces (IRS), SWIPT, and heterogeneous cellular networks (HCNs) [15]. In SWIPT-enabled cooperative NOMA, the far user's performance gain is achieved via assistance from a near user–albeit under the ideal SIC assumption [112]. Finally, embedding NOMA within HCNs—by deploying it at low-power picocell base stations—addresses the twin goals of supporting high user density and elevated data rates [113, 114]. HCNs can further accommodate surging data demands through load-balancing across multiple tiers and NOMA-enhanced 6G standards. To mitigate cross-tier interference, which exacerbates outage probability, switching-based hybrid mmWave/UHF architectures have been proposed [115]. Additionally, cooperative CRN models within HCNs enable in-band bidirectional D2D communications, where RF signals from cognitive nodes simultaneously convey information and power via SWIPT [116].

The power amplifier (PA) is a critical element of the base-station transmitter. By equipping BSs with high-power amplifiers (HPAs), networks can support more devices and heavier traffic loads. The additional transmit power improves the link budget—the ratio of received signal power to the minimum required for reliable communication—thereby enhancing connection reliability, especially at millimeter-wave frequencies where path loss is severe [117]. HPAs also extend coverage and enlarge handover zones, smoothing handovers and reducing call drops [118]. However, operating an HPA near saturation to maximize efficiency introduces non-linear distortion (NLD) and incurs higher power consumption, heat dissipation, RF interference, and linearity challenges [119]. This NLD degrades performance across different network topologies [2], and is expected to become more pronounced in 6G systems operating at even higher millimeter-wave or terahertz bands. To quantify these effects, relays have been modeled with soft envelope limiters (SEL), solid-state power amplifiers (SSPAs), and traveling-wave tube amplifiers (TWTAs) [120], and the impact on time-switching SWIPT-enabled amplify-and-forward two-way relay networks over Nakagami-m fading channels has been analyzed [121]. Decoding performance in NOMA hinges on SIC [122, 123]. After the high-SNR symbol is decoded, SIC removes it from the composite signal to enable subsequent detections. SIC effectiveness depends on optimal power allocation, precise channel-state information (CSI), and accurate amplitude/phase prediction for each symbol [124]. For tractable analysis and bound derivation, we assume perfect CSI, consistent with foundational studies on NOMA in multi-tier HCNs [112, 114]. In practice, CSI imperfections (e.g., estimation or quantization errors) lead to incomplete subtraction—imperfect SIC (i-SIC)—which leaves residual interference at the low-SNR user and must be included in performance evaluations. The outage probability of cooperative NOMA with i-SIC in underlay cognitive radio networks has been characterized under Gaussian SIC errors [125], and a distributed clustering/resource-allocation framework for i-SIC-based NOMA in HCNs was proposed [126]. The impact of i-SIC on energy-harvesting multi-relay systems is studied in [127]; the coordinated direct-and-relay transmission (CDRT) protocol with incremental signaling and i-SIC in downlink NOMA-IoT networks appears in [128]; SWIPT-enabled NOMA architectures integrating i-SIC for small-area IoT uplink/downlink are detailed in [129]; and a hybrid HCN employing mmWave and sub-6 GHz links under NOMA/i-SIC assumptions is presented in [130].

2.1 Motivation and Contribution

This chapter focuses on the impact of non-linear distortions introduced by HPAs in an HCN employing NOMA, specifically under the influence of i-SIC. These NLDs arise when HPAs are operated near saturation to enhance power efficiency, which is a practical limitation in real-world deployments. The following are the key contributions made in this chapter:

- To enhance energy efficiency, power amplifiers (PAs) are typically operated near their saturation region, which leads to significant NLD. These distortions deteriorate the quality of the transmitted signal, thereby increasing the outage probability of wireless network users [2]. Prior studies such as [114] have assessed HCN-NOMA performance without accounting for this critical non-linearity. To address this gap, we investigate the HCN-NOMA system's outage performance under the influence of HPA-induced NLD. Our analysis compares the outage probability of users across different network tiers by incorporating the effect of HPA non-linearity. To emulate realistic deployment scenarios, we adopt a PPP framework for modeling the spatial distribution of MBSs and PBSs. The use of PPP is essential as it provides a mathematically tractable and spatially consistent model for HCNs, enabling accurate and practical performance evaluation under real-world network geometries.

- SIC is a key enabler in NOMA systems for decoding superimposed signals. However, due to hardware imperfections and non-ideal channel estimation, SIC can be imperfect (i-SIC), leaving behind residual interference denoted by (β) [123, 131]. This residual interference can significantly affect system performance, especially at lower SNR regimes. In this context, we investigate the performance of HCN-NOMA systems considering both perfect and imperfect SIC and benchmark them against OMA systems. Our analysis evaluates user performance at different tiers under varying levels of residual interference and transmit SNR, highlighting the degradation in outage probability due to i-SIC. This investigation underscores the necessity of considering i-SIC in practical NOMA-based HCN systems.
- The non-linear behavior of HPA and residual interference from i-SIC are treated as independent impairments. However, in this work, we jointly examine these impairments to reflect more realistic system conditions. An analytical framework is formulated to assess the compound effects of HPA-induced non-linearity and i-SIC in an HCN environment with PPP-distributed users. Outage probability expressions are derived for both perfect and imperfect SIC cases in the presence of non-linear HPA distortions. These expressions are presented and discussed in Sect. 2.3.1. This joint analysis allows for a more comprehensive understanding of how multiple hardware-induced distortions interact and influence system performance.
- To support efficient performance evaluation and system design, we derive approximate closed-form expressions for outage probability. These expressions provide a simplified yet accurate means of analyzing both linear and non-linear systems, with and without NOMA. By capturing the essence of system behavior, the closed-form results allow for rapid assessment of the effects of key system parameters on performance. Additionally, we derive closed-form outage expressions for a practical, non-linear NOMA system under both perfect and imperfect SIC conditions. This generalized framework used to evaluate the performance of HCN-NOMA networks under realistic non-idealities.
- We provide a detailed quantitative comparison of spectral efficiency (SE) and energy efficiency (EE) across various network configurations and impairment conditions. These performance metrics are illustrated through comprehensive graphical results, allowing for intuitive visualization of trade-offs and performance trends. Furthermore, extensive Monte Carlo simulations are conducted to validate the derived analytical results. The strong agreement between simulation and analytical findings confirms the accuracy and robustness of the proposed models and expressions.

2.2 System Model

The system under consideration consists of a two-tier HCN comprising MBS and PBS, as depicted in Fig. 2.1. The PBS tier operates using NOMA, which enables simultaneous transmission to users with heterogeneous channel conditions and diverse QoS requirements over the same frequency resource in the downlink. The spatial deployment of MBSs and PBSs is modeled using a homogeneous PPP, which provides a mathematically tractable and spatially realistic representation of node distributions in wireless networks. This modeling approach facilitates analytical evaluation of network performance under practical deployment scenarios.

BSs and users are independently distributed following PPPs, where the set of BS locations in the tth tier is denoted by Ω_t, and the corresponding BS density is given by λ_t, with $t \in \{m, p\}$ representing macro and pico tiers, respectively. Similarly, users are distributed according to a PPP with density λ_u, and their spatial distribution is denoted as Ω_u. Each BS in the tth tier transmits with power P_t and serves users within a coverage radius $\mathcal{Y}_t$. For the NOMA-enabled PBS tier, the power allocation for the nth user is represented by the coefficient a_n.

To establish user association, the nearest neighbor connection rule is applied such that each macro user (MU) connects to its closest MBS. The signal attenuation due to path loss is modeled using a bounded path loss function $P(r) = \frac{1}{1+r_t^{\nu_t}}$ [132], where r_t is the distance between the serving BS and the typical user in the tth tier, and ν_t is the path loss exponent. This model ensures that the path loss remains bounded and does not exceed unity even at short distances. Accordingly, the effective channel power gain for the typical user associated with the tth tier is expressed as $|h_t|^2 = |\hat{h}_t|^2 P(r)$, where $\hat{h}_t$ denotes the small-scale fading component, modeled as a Rayleigh distributed random variable. The system operates over a normalized transmission bandwidth 1 Hz. Let R represent the desired data rate of the typical user; successful communication is assumed when the instantaneous transmission rate exceeds this target rate. In this analysis, the unordered channel gain is denoted by $|\tilde{h}_t|^2$, whereas the ordered channel gain is represented by $|h_t|^2$, both pertaining to the tth tier under NOMA transmission.

Having established the system model and its key components, we now proceed to evaluate the system's performance. A fundamental measure in this regard is the SINR, which provides critical insights into the effectiveness of the proposed scheme. Since the detailed formulation and significance of SINR have already been discussed in Chap. 1, Sect. 1.8, we directly present its expressions for the proposed system to facilitate performance comparison.

2.2.1 SINR Experienced at a Typical MBS-Connected User

The MBS transmits the signal $X_{m,tx} = \sqrt{P_m} x_m$, and the typical MU receives

$$X_{m,rx} = \underbrace{\sqrt{P_m} x_m \tilde{h}_{m_o}}_{\text{Intended desired signal}} + \underbrace{\sum_{t\in(m,p)} \sqrt{P_t} x_t \sum_{r\in\Omega_t\setminus\{m_o\}} \tilde{h}_r}_{\text{Co-tier and Cross-tier interference by BSs}} + n_m, \tag{2.1}$$

where x_m is the information-bearing signal for the MU, and n_m corresponds to the channel noise. Accordingly, the SINR at a typical MU is formulated as [114]

$$m = \frac{\rho_m P_m |\tilde{h}_{m_o}|^2}{\sum\limits_t \rho_t^I \Upsilon_t + 1}, \tag{2.2}$$

where the transmit SNR at the MBS tier, denoted by ρ_m, is defined as $\rho_m = \mathbb{E}\left[x_m^2\right]/\sigma_m^2$, where σ_m^2 indicates the variance of the additive noise. Here, P_t refers to the transmit power associated with the tth tier, and $\mathcal{Y}_t$ designates its coverage region. The transmit SNR corresponding to interfering base stations (BSs) from the tth tier is given by $\rho_t^I = P_t/\sigma_m^2$, and the aggregate interference from that tier is captured by the product $\rho_t^I \Upsilon_t$. Leveraging Slivnyak's theorem [133, 134], the analysis assumes that the typical MU is positioned at the origin and the serving MBS is located at point m_o. Under this setup, the co-tier interference from other MBSs is quantified as $\rho_m^I \Upsilon_m$, where $\Upsilon_m = \sum_{i \in \Omega_m \setminus \{m_o\}} |\tilde{h}_i|^2$, with $|\tilde{h}_i|^2$ representing the total channel gain from the ith interfering MBS to the typical MU. Meanwhile, the term $|\tilde{h}_{m_o}|^2$ denotes the overall unordered channel gain between the tagged MBS and the MU. In a similar manner, cross-tier interference originating from the PBS tier is modeled as $\rho_p^I \Upsilon_p$, where $\Upsilon_p = \sum_{j \in \Omega_p} |\tilde{h}_j|^2$, and $|\tilde{h}_i|^2$ denotes the composite channel gain from the jth PBS to the MU.

2.2.2 SINR Observed at the PU Under OMA

As detailed in Sect. 2.2.1, the SINR for a user connected to the PBS tier (PU) is formulated accordingly.

$$\gamma_p = \frac{\rho_p P_p |\tilde{h}_{p_o}|^2}{\sum\limits_t \rho_t^I \Upsilon_t + 1}, \tag{2.3}$$

where the transmit SNR at the PBS, denoted by ρ_p, is defined as $\rho_p = \mathbb{E}\left[x_p^2\right]/\sigma_p^2$, where x_p represents the desired signal intended for the typical PU, and σ_p^2 is the corresponding noise variance. The transmit SNR from the tth interfering tier is expressed as $\rho_t^I = P_t/\sigma_p^2$. The typical PU is subject to co-tier interference from other PBSs in its tier, modeled as $\rho_p^I \Upsilon_p$. Assuming, without loss of generality, that the tagged PBS is located at p_o and the PU is situated at the origin—as justified by Slivnyak's theorem [133, 134]—the co-tier interference from the PBS tier is given by $\rho_p^I \Upsilon_p$, where $\Upsilon_p = \sum_{j \in \Omega_p \setminus \{p_o\}} |\tilde{h}_j|^2$, and $|\tilde{h}_j|^2$ denotes the overall channel gain from the jth PBS to the PU. In addition, the PU also experiences cross-tier interference from the MBS tier, expressed as $\rho_m^I \Upsilon_m$, where $\Upsilon_m = \sum_{i \in \Omega_m} |\tilde{h}_i|^2$, with $|\tilde{h}_i|^2$ representing the overall channel gain from the ith MBS to the same PU.

2.2.3 SINR at the PU Under NOMA with Perfect SIC

Consider a scenario where the PBS employs NOMA to simultaneously serve M_p users. The users are ordered based on their channel gains such that $|h_1^{p_o}|^2 \leq \cdots \leq |h_{M_p}^{p_o}|^2$. In accordance with the NOMA principle in the power domain, the power allocation coefficients are assigned as $a_1 \geq \cdots \geq a_{M_p}$, with users having weaker channels receiving more power. Common misconceptions surrounding NOMA power allocation and related issues are thoroughly addressed in [135]. Let x_i denote the intended signal for the ith user, assuming equal signal power levels across all users. The composite signal transmitted by the PBS is expressed as $X_{p,tx} = \sum_{i=1}^{M_p} x_i \sqrt{a_i P_p}$. Consequently, the received signal at the kth user is given by

$$X_{p,rx} = h_k^{p_o} \left(\sum_{i=1}^{M_p} x_i \sqrt{a_i P_p} \right) + \underbrace{\sum_{t \in (m,p)} \sqrt{P_t} x_t \sum_{r \in \Omega_t \setminus \{p_o\}} h_r}_{\text{Co-tier and Cross-tier interference by BSs}} + n_k. \tag{2.4}$$

To enable SIC, the kth user attempts to decode the signal intended for the jth user. The resulting SINR at the kth user can be written as

$$\gamma_{j \to k}^p = \frac{\rho_p P_p a_j \left| h_k^{p_o} \right|^2}{\rho_p P_p \left| h_k^{p_o} \right|^2 \sum\limits_{l=j+1}^{M_p} a_l + \sum\limits_t \rho_t^I \Upsilon_t + 1}, \tag{2.5}$$

where the term a_n refers to the power allocation associated with user n, where $n = \{k, j, l\}$. The kth user decodes its own message for which the SINR is given by

$$\gamma_k^p = \frac{\rho_p P_p a_k \left| h_k^{p_o} \right|^2}{\rho_p P_p \left| h_k^{p_o} \right|^2 \sum\limits_{l=k+1}^{M_p} a_l + \sum\limits_t \rho_t^I \Upsilon_t + 1}. \tag{2.6}$$

2.2.4 SINR at the PU Considering Imperfect SIC in NOMA

In the case of i-SIC, the high-SNR signal is not fully decoded by users with stronger channel conditions. Consequently, the subtraction of this signal from the superimposed NOMA transmission becomes inaccurate, leaving a residual component. This residual interferes with the decoding process of users with better channel gains, thereby degrading performance. To perform SIC, the kth user attempts to decode the message intended for the jth user. Accordingly, the SINR at the kth user is given by

$$\gamma_{j\to k}^{p,iSIC} = \frac{\rho_p P_p a_j \left|h_k^{p_o}\right|^2}{\rho_p P_p \left|h_k^{p_o}\right|^2 \left[\sum\limits_{l=j+1}^{M_p} a_l + \sum\limits_{\delta=1}^{j-1} \beta_\delta a_\delta\right] + \sum\limits_t \rho_t^I \Upsilon_t + 1}. \tag{2.7}$$

In this context, a_n indicates the power allocation coefficient assigned to the user indexed by $n = \{k, j, l\}$. The term $\beta_\delta = \beta; \forall\delta \in (1, 2, ..., M_p - 1)$, captures the degree of residual interference encountered by users due to i-SIC. Specifically, β reflects the proportion of high-SNR signal leakage that persists and interferes with the decoding process of lower-SNR users. The value of β lies within the range $0 \leq \beta \leq 1$: $\beta = 0$ denotes an ideal SIC scenario with complete cancellation of stronger signals, whereas $\beta = 1$ implies the absence of any interference cancellation, effectively disabling SIC. For values where $0 < \beta < 1$, SIC is considered only partially effective, resulting in a residual fraction of the stronger user's signal interfering with the decoding process. It is worth noting that the user with the weakest channel—typically the one located farthest from the PBS—does not apply SIC and instead treats all other users' signals as noise; thus, its reception remains unaffected by the imperfections of SIC. While a more comprehensive modeling of imperfect SIC could incorporate probabilistic decoding failures and residuals modeled as additional Gaussian noise components, such scenarios—explored in works like [125, 136]—are beyond the scope of this study. Here, we adopt the deterministic i-SIC model proposed in [131] for analytical tractability.

Similarly, the kth user decodes its own message, SINR is given by

$$\gamma_k^{p,iSIC} = \frac{\rho_p P_p a_k \left|h_k^{p_o}\right|^2}{\rho_p P_p \left|h_k^{p_o}\right|^2 \left[\sum\limits_{l=k+1}^{M_p} a_l + \sum\limits_{\delta=1}^{k-1} \beta_\delta a_\delta\right] + \sum\limits_t \rho_t^I \Upsilon_t + 1}. \tag{2.8}$$

2.2.5 SINR at the PU in NOMA Under Non-linear HPAs

The transmit signal from the PBS can be expressed as $X_{tx}^p = \sum_{i=1}^{M_p} x_i \sqrt{a_i P_p}$. Consequently, the received signal at the kth user is represented by (2.4). To characterize the non-linear behavior of HPAs at the base station, the polynomial model is widely adopted due to its analytical tractability [137]. Under this framework, the output of a memoryless non-linear HPA is modeled as follows:

$$z_{out} = z_{in} \sum_{n=0}^{N} \alpha_{2n+1} |z_{in}|^{2n}, \tag{2.9}$$

where z_{in} denotes the input signal to the memoryless non-linear system. When incorporating the effects of HPA non-linearity into NOMA transmission, the input z_{in} in (2.9) is substituted with . The coefficients α_{2n+1} represent the non-linear terms of the polynomial model, with the polynomial order denoted by N. Accordingly, the received signal at the kth user can be expressed as follows:

$$X_{k,rx}^{hpa,p} = h_k^{po}\left(\sqrt{a_k P_p x_k}\sum_{n=0}^{N}\alpha_{2n+1}\left|\sum_{i=1}^{M_p}\sqrt{a_i P_p x_i}\right|^{2n} + \sum_{\substack{u=1\\u\neq k}}^{M_p}\sqrt{a_u P_p x_u}\sum_{n=0}^{N}\alpha_{2n+1}\left|\sum_{i=1}^{M_p}\sqrt{a_i P_p x_i}\right|^{2n}\right)$$
$$+ \underbrace{\sum_{t\in(m,p)}\sqrt{P_t x_t}\sum_{r\in\Omega_t\setminus\{po\}} h_r}_{\text{Co-tier and Cross-tier interference by BSs}} + n_k. \tag{2.10}$$

The signal received at the kth user from the BS, as defined in (2.10), can alternatively be expressed as

$$X_{k,rx}^{hpa,p} = h_k^{p_o}\sqrt{P_p}s_k + \underbrace{\sum_{t\in(m,p)}\sqrt{P_t x_t}\sum_{r\in\Omega_t\setminus\{p_o\}} h_r}_{\text{Co-tier and Cross-tier interference by BSs}} + n_k, \tag{2.11}$$

where s_k includes the following components:

$$s_k = \underbrace{\alpha_1\sqrt{a_k}}_{\text{Linear-gain}}\, x_k + \underbrace{\sqrt{a_k}x_k\sum_{n=1}^{N}\alpha_{2n+1}\left|\sum_{i=1}^{M_p}\sqrt{a_i P_p x_i}\right|^{2n}}_{\text{Non-Linear distortion}}$$
$$+ \underbrace{\sum_{u=k+1}^{M_p}\sqrt{a_u}x_u\sum_{n=0}^{N}\alpha_{2n+1}\left|\sum_{i=1}^{M_p}\sqrt{a_i P_p x_i}\right|^{2n}}_{\text{Inter-User Interference}}$$
$$+ \underbrace{\sum_{w=1}^{k-1}\sqrt{a_w}x_w\sum_{n=0}^{N}\alpha_{2n+1}\left|\sum_{i=1}^{M_p}\sqrt{a_i P_p x_i}\right|^{2n}}_{\text{SIC}}. \tag{2.12}$$

For the sake of simplicity and without loss of generality, a third-order polynomial model (i.e., $N = 1$) is adopted to characterize the HPA non-linearity, which serves as a reliable approximation [138]. The SINR at the kth user for decoding the message intended for the jth user, denoted by $\gamma_{j\to k}^{hpa,p}$ with $j \leq k$, is expressed as follows:

$$\gamma_{j\to k}^{hpa,p} = \frac{\rho_p P_p a_j \left|h_k^{p_o}\right|^2 |\alpha_1|^2}{\rho_p P_p \left|h_k^{p_o}\right|^2\left[\sum\limits_{l=j+1}^{M_p}\mu + \epsilon\right] + \sum\limits_t \rho_t^{\mathcal{I}}\Upsilon_t + 1}, \tag{2.13}$$

where $\mu = a_l\left(|\alpha_1|^2 + |\alpha_3|^2 P_p^2\left|\sum_{i=1}^{M_p}\sqrt{a_i}\right|^4\right)$ and $\epsilon = |\alpha_3|^2 P_p^2 a_j\left|\sum_{i=1}^{M_p}\sqrt{a_i}\right|^4$. In a similar manner, the SINR at the kth user for decoding its own message can be represented as

$$\gamma_k^{hpa,p} = \frac{\rho_p P_p a_k \left|h_k^{p_o}\right|^2 |\alpha_1|^2}{\rho_p P_p \left|h_k^{p_o}\right|^2 \left[\sum\limits_{l=k+1}^{M_p} \mu + \epsilon\right] + \sum\limits_t \rho_t^I \Upsilon_t + 1}, \tag{2.14}$$

where $\mu = a_l \left(|\alpha_1|^2 + |\alpha_3|^2 P_p^2 \left|\sum_{i=1}^{M_p} \sqrt{a_i}\right|^4 \right)$ and $\epsilon = |\alpha_3|^2 P_p^2 a_k \left|\sum_{i=1}^{M_p} \sqrt{a_i}\right|^4$.

2.2.6 SINR at the PU Under NOMA with Imperfect SIC and Non-linear HPA Effects

The non-linear signal received at the kth user is described by (2.10) and (2.11), where s_k encompasses various components, including the residual interference resulting from imperfect SIC at the kth user,

$$\begin{aligned} s_k = {} & \underbrace{\alpha_1 \sqrt{a_k}}_{\text{Linear-gain}}\, x_k + \underbrace{\sqrt{a_k} x_k \sum_{n=1}^{N} \alpha_{2n+1} \left| \sum_{i=1}^{M_p} \sqrt{a_i P_p} x_i \right|^{2n}}_{\text{Non-Linear distortion}} \\ & + \underbrace{\sum_{u=k+1}^{M_p} \sqrt{a_u} x_u \sum_{n=0}^{N} \alpha_{2n+1} \left| \sum_{i=1}^{M_p} \sqrt{a_i P_p} x_i \right|^{2n}}_{\text{Inter-User Interference}} \\ & + \underbrace{\sum_{w=1}^{k-1} \sqrt{\beta_w} \sqrt{a_w} x_w \sum_{n=0}^{N} \alpha_{2n+1} \left| \sum_{i=1}^{M_p} \sqrt{a_i P_p} x_i \right|^{2n}}_{\text{iSIC}}. \end{aligned} \tag{2.15}$$

Applying the equations above considering the 3rd-order polynomial model for HPA, we can express the SINR at the kth user to decode the jth user message given by $\gamma_{j\to k,iSIC}^{hpa,p}$, when $j \leq k$, shown in (2.16)

$$\begin{aligned} \gamma_{j\to k,iSIC}^{hpa,p} = {} & \left[\rho_p P_p a_j \left|h_k^{p_o}\right|^2 |\alpha_1|^2\right] \\ & \times \left[\rho_p P_p \left|h_k^{p_o}\right|^2 \left\{ \sum_{\substack{l=1 \\ l\neq k}}^{M_p} a_l \beta_l \left(|\alpha_1|^2 + |\alpha_3|^2 P_p^2 \left| \sum_{i=1}^{M_p} \sqrt{a_i} \right|^4 \right) + |\alpha_3|^2 P_p^2 a_j \left| \sum_{i=1}^{M_p} \sqrt{a_i} \right|^4 \right\} + \sum_t \rho_t^I \Upsilon_t + 1 \right]^{-1} \end{aligned} \tag{2.16}$$

where

$$\beta_l = \begin{cases} 1, & \text{if } l > j \\ \beta, & \text{otherwise.} \end{cases} \tag{2.17}$$

The SINR corresponding to the kth user's self-message decoding can be written as (2.18)

$$\gamma_{k,iSIC}^{hpa,p} = \left[\rho_p P_p a_k \left|h_k^{p_o}\right|^2 |\alpha_1|^2\right] \times \left[\rho_p P_p \left|h_k^{p_o}\right|^2 \left\{\sum_{\substack{l=1 \\ l \neq k}}^{M_p} a_l \beta_l \left(|\alpha_1|^2 + |\alpha_3|^2 P_p^2 \left|\sum_{i=1}^{M_p} \sqrt{a_i}\right|^4\right) + |\alpha_3|^2 P_p^2 a_k \left|\sum_{i=1}^{M_p} \sqrt{a_i}\right|^4\right\} + \sum_t \rho_t^I \Upsilon_t + 1\right]^{-1} \tag{2.18}$$

where

$$\beta_l = \begin{cases} 1, & \text{if } l > k \\ \beta, & \text{otherwise.} \end{cases} \tag{2.19}$$

2.3 Performance Analysis

This section evaluates the rate outage probability across various tiers for the systems described in Sect. 2.2. Generally, outage probability quantifies the chance that a wireless communication link undergoes considerable performance degradation—manifested as a decline in signal quality, data rate, or a rise in error occurrences. Minimizing outage probability is critical for enhancing the reliability and overall performance of wireless networks. An outage condition is met if the instantaneous rate is less than the user's desired data rate, detailed in Chap. 1, Sect. 1.8.

2.3.1 Rate Outage Analysis

Outage Analysis for MBS Tier

The outage probability for an OMA user connected to the MBS is presented in [130] as

$$\begin{aligned} \mathcal{P}_O^m &= \mathcal{P}\left(\omega_m \times \log_2(1+\gamma_m) < R\right), \\ &= \mathcal{P}\left(|\tilde{h}_{m_o}|^2 < \frac{\phi}{\rho_m P_m}\left(1 + \sum_t \rho_t^I \Upsilon_t\right)\right). \end{aligned} \tag{2.20}$$

In this context, ω_m represents the portion of the total bandwidth allocated to the typical MU under OMA. The SINR at the MU, denoted by $_m$, is given by (2.2). The user's target data rate, R, sets the SINR threshold, which is calculated as $\phi = 2^{2R} - 1$.

Proposition 2.1 *Under the homogeneous PPP assumption and employing polar coordinates, the CDF of the unordered channel gain for the MBS tier can be formulated as follows [114]:*

$$F_{|\tilde{h}_m|^2}(y) = 2\pi\lambda_m \int_0^{\mathcal{Y}_m} (1 - e^{-(1+r_m^{\nu_m})y}) e^{-\pi\lambda_m r_m^2} r_m dr_m. \tag{2.21}$$

By applying Gauss–Chebyshev quadrature [139], Eq. (2.21) can be approximated as,

$$F_{|\tilde{h}_m|^2}(y) \simeq \pi\lambda_m \mathcal{Y}_m^2 \sum_{\iota=0}^{I} b_\iota^m e^{-c_\iota^m y}. \tag{2.22}$$

Proof A detailed explanation is available in Appendix A.1. □

Thus, (2.20) can be rewritten to represent the outage probability as follows:

$$\begin{aligned}\mathcal{P}_O^m &\approx \pi\lambda_m \mathcal{Y}_m^2 \sum_{\iota=0}^{I} b_\iota^m e^{-c_\iota^m \frac{\phi}{\rho_m P_m}(1+\sum_t \rho_t^I \Upsilon_t)}, \\ &\approx \pi\lambda_m \mathcal{Y}_m^2 \sum_{\iota=0}^{I} b_\iota^m e^{-c_\iota^m \frac{\phi}{\rho_m P_m}} \prod_t \mathcal{L}_{\Upsilon_t}\left(s_m \rho_t^I\right),\end{aligned} \tag{2.23}$$

where I provides a trade off between complexity and accuracy, $b_\iota^m = -w_I\sqrt{1-\theta_\iota^2}$ $\left(\frac{1}{2}(\theta_\iota+1)\right) e^{-\pi\lambda_m\left(\frac{1}{2}(\theta_\iota+1)\mathcal{Y}_m\right)^2}$, $w_I = \frac{\pi}{I}$, $\theta_\iota = \cos\left(\frac{2\iota-1}{2I}\pi\right)$, $b_0 = -\sum_{\iota=1}^{I} b_\iota^m$, $c_\iota^t = 1 + \left(\frac{\mathcal{Y}_t}{2}\theta_\iota + \frac{\mathcal{Y}_t}{2}\right)_t^\nu$, $c_0 = 0$, $s_t = \frac{c_\iota^t \phi}{\rho_t P_t}$ such that $t \in \{m, p\}$.

The Laplace transform (LT) of the interference originating from the tth tier is denoted by $\mathcal{L}_{\Upsilon_t}(s)$, and is expressed as follows [132]:

$$\mathcal{L}_{\Upsilon_t}(s) = e^{\pi\lambda_t\left(s^{\xi_t}\Gamma(1-\xi_t,s) - s^{\xi_t}\Gamma(1-\xi_t)\right)}, \tag{2.24}$$

where $\xi_t = 2/\nu_t$, $\Gamma(z) = \int_0^\infty x^{z-1}e^{-x}dx$ and $\Gamma(z,x) = \int_x^\infty t^{z-1}e^{-t}dt$.

Outage Performance Evaluation of the PBS Tier Under OMA

Under the assumption of a uniform distance between the PBS and the typical PU, the rate outage probability is given as:

$$\begin{aligned}\mathcal{P}_O^p &= \mathcal{P}\left(\omega_p \times \log_2(1+\gamma_p) < R\right), \\ &= \mathcal{P}\left(|\tilde{h}_{p_o}|^2 < \frac{\phi}{\rho_p P_p}\left(1 + \sum_t \rho_t^I \Upsilon_t\right)\right),\end{aligned} \tag{2.25}$$

where ω_p denotes the portion of bandwidth assigned to the OMA-based typical PU, and $_p$ represents the SINR at the PU as defined in (2.3). The CDF of the unordered channel gain for the PBS tier, derived using polar coordinates under the assumption of a homogeneous PPP, is given as [140]:

$$F_{|\widetilde{h}_{po}|^2}(y) = \frac{2}{\mathcal{Y}_p^2}\int_0^{\mathcal{Y}_p}(1 - e^{-(1+z^{\nu_p})y})zdz. \tag{2.26}$$

By applying the G-C quadrature [139] to (2.26), we get

$$F_{|\widetilde{h}_{po}|^2}(y) \simeq \sum_{\iota=0}^{I} b_\iota^p e^{-c_\iota^p y}. \tag{2.27}$$

Thus, the expression for outage probability in (2.25) can be re-expressed as:

$$\begin{aligned}\mathcal{P}_O^p &\approx \sum_{\iota=0}^{I} b_\iota^p e^{-c_\iota^p \frac{\phi}{\rho_p P_p}(1+\sum_t \rho_t^I \Upsilon_t)},\\ &\approx \sum_{\iota=0}^{I} b_\iota^p e^{-c_\iota^p \frac{\phi}{\rho_p P_p}} \prod_t \mathcal{L}_{\Upsilon_t}(\rho_t^I s_p),\end{aligned} \tag{2.28}$$

where $b_\iota^p = -w_I\sqrt{1-\theta_\iota^2}\left(\frac{1}{2}(\theta_\iota + 1)\right), w_I = \frac{\pi}{I}, \theta_\iota = \cos\left(\frac{2\iota-1}{2I}\pi\right), b_0 = -\sum_{\iota=1}^{I} b_\iota^p$.

Outage Probability Analysis of PBS Tier with NOMA Considering Imperfect SIC and HPA Non-Linearity

In scenarios where a predefined target rate R is imposed by the user's QoS requirements, the outage probability is formulated as:

$$\mathcal{P}_k^{O,p} = \mathcal{P}(\gamma_{k,iSIC}^{hpa,p} < \phi_k, \gamma_{j\to k,iSIC}^{hpa,p} < \phi_j), \tag{2.29}$$

where $\gamma_{j\to k,iSIC}^{hpa,p}$ and $\gamma_{k,iSIC}^{hpa,p}$ are given in (2.16), and (2.18) respectively. As the outage probability depends on both successful SIC and correct decoding of the user's own message, it can be expressed at the kth user as:

$$\mathcal{P}_k^{O,p} = \mathcal{P}\left(|h_k^{p_o}|^2 < \frac{\psi_{max,iSIC}^{NL}(1+\sum_t \rho_t^I \Upsilon_t)}{\rho_p P_p}\right). \tag{2.30}$$

This gives the outage probability as

$$\mathcal{P}_k^{O,p} = F_{|h_k^{p_o}|^2}(y), \tag{2.31}$$

where $y = \frac{\psi_{max,iSIC}^{NL}(1+\sum_t \rho_t^I \Upsilon_t)}{\rho_p P_p}$ and $\psi_{max,iSIC}^{NL} = \max\left(\psi_{(1,iSIC)}^{NL}, \psi_{(2,iSIC)}^{NL}, \ldots, \psi_{(k,iSIC)}^{NL}\right)$. For a user with $k = 1$, outage probability is simply calculated with $y = \frac{\psi_{j,iSIC}^{NL}(1+\sum_t \rho_t^I \Upsilon_t)}{\rho_p P_p}$, since, it decodes only its own message and does not perform SIC. $\psi_{j,iSIC}^{NL}$ is given below,

$$\psi_{j,iSIC}^{NL} = \frac{\phi_j}{a_j|\alpha_1|^2 - \phi_j\left[\sum_{\substack{l=1\\l\neq j}}^{M_p} a_l\beta_l\left(|\alpha_1|^2 + |\alpha_3|^2 P_p^2 |\sum_{i=1}^{M_p}\sqrt{a_i}|^4\right) + |\alpha_3|^2 P_p^2 a_j |\sum_{i=1}^{M_p}\sqrt{a_i}|^4\right]}. \tag{2.32}$$

where ϕ_j is defined as $2^{R_j} - 1$, with R_j specifying the target data rate for the jth user, and the residual interference parameter is taken as in (2.17).

Proposition 2.2 *As outlined in [141], the expression for the ordered channel gain in terms of the unordered channel gain for the PBS tier is* $F_{|\tilde{h}_{po}|^2}(y)$*, given as:*

$$F_{|h_k^{po}|^2}(y) = \epsilon_k \sum_{r=0}^{M_p-k} \binom{M_p-k}{r} \frac{(-1)^r}{k+r}\left(F_{|\tilde{h}_{po}|^2}(y)\right)^{r+k}. \tag{2.33}$$

Proof A detailed explanation is available in Appendix A.2. □

Applying the multinomial theorem to (2.33) after substituting (2.27) yields the CDF of the ordered channel gains as:

$$F_{|h_k^{po}|^2}(y) \simeq \epsilon_k \sum_{r=0}^{M_p-k} \binom{M_p-k}{r} \frac{(-1)^r}{k+r} \sum_{T_k^r} \binom{k+r}{q_0+\ldots+q_N} \times \prod_{\iota=0}^{I} (b_\iota^p)^{q_\iota} e^{-\sum_{\iota=0}^{I} q_\iota c_\iota^p y}. \tag{2.34}$$

Assuming the typical PU is uniformly distributed around the PBS and considering the ordered channel gains, the outage probability corresponding to the k-th typical PU is given by:

$$\mathcal{P}_k^{O,p} \approx \epsilon_k \sum_{r=0}^{M_p-k} \binom{M_p-k}{r} \frac{(-1)^r}{k+r} \sum_{T_k^r} \binom{k+r}{q_0+\ldots+q_N}$$
$$\times \prod_{\iota=0}^{I} (b_\iota^p)^{q_\iota} e^{-\sum_{\iota=0}^{I} q_\iota c_\iota^p \frac{\psi_{\max,iSIC}^{NL}}{\rho_p P_p}} \prod_t \mathcal{L}_{\Upsilon_t}(\rho_t^I s_p), \tag{2.35}$$

where $\epsilon_k = \frac{M_p!}{(k-1)!(M_p-k)!}$, $s_p = \frac{\psi_{\max,iSIC}^{NL} \sum_{\iota=0}^{I} q_\iota c_\iota^p}{\rho_p P_p}$, $T_k^r = \left\{(q_0, \ldots, q_N) | \sum_{i=0}^{N} q_i = k+r\right\}$, $\binom{k+r}{q_0+\ldots+q_N} = \frac{(k+r)!}{q_0!\ldots q_N!}$.

It can be noted that the user with $k = 1$ does not perform SIC, hence the term $\psi_{\max,iSIC}^{NL} = \psi_{1,iSIC}^{NL}$.

Note:

- In the case of a linear HPA at the PBS tier, the outage probability expression for the kth user retains the structure of (2.35), except that $\psi_{j,iSIC}^{NL}$ is replaced by $\psi_j = \phi_j \times (a_j - \phi_j \sum_{i=j+1}^{M_p} a_i)^{-1}$.

- For the scenario involving a non-linear HPA and no imperfect SIC at the PBS tier, the rate outage probability at the k-th user is analogous to that in (2.35), where $\psi_{j,iSIC}^{NL}$ is replaced by ψ_j^{NL} as given below,

$$\psi_j^{NL} = \frac{\phi_j}{a_j\,|\alpha_1|^2 - \phi_j\left[\sum_{l=j+1}^{M_p} a_l\left(|\alpha_1|^2 + |\alpha_3|^2 P_p^2 |\sum_{i=1}^{M_p}\sqrt{a_i}|^4\right) + |\alpha_3|^2 P_p^2 a_j |\sum_{i=1}^{M_p}\sqrt{a_i}|^4\right]}. \tag{2.36}$$

2.3.2 System Throughput Analysis

Following the separate analysis of outage probabilities for CCU and CEU in the previous section, the system's aggregate performance is evaluated using the system throughput metric (bps/Hz), as outlined in [112].

$$\tau_{sys} = (1 - \mathcal{P}_{CEU}^{O,p})R_1 + (1 - \mathcal{P}_{CCU}^{O,p})R_2. \tag{2.37}$$

In (2.35), setting $k = 1$ and $k = 2$ yields the outage probabilities $\mathcal{P}_{CEU}^{O,p}$ and $\mathcal{P}_{CCU}^{O,p}$ at the PBS tier for CEU and CCU, respectively, as per [142].

2.3.3 Energy Efficiency Analysis

Energy efficiency serves as a key performance indicator that evaluates how effectively a system maximizes data throughput while minimizing energy consumption per bit of information transmitted. A higher energy efficiency value implies improved throughput with reduced energy consumption, contributing to a more cost-effective communication framework. In energy-efficient NOMA system designs, the goal is to strike a balance between achieving high data rates and conserving power. We compare the energy efficiency performance of both linear and non-linear OMA with their NOMA counterparts—considering cases with and without i-SIC. The system's energy efficiency can be defined as the ratio of the total transmitted data to the total energy consumed, as discussed in [130, 143]. The total data delivered is derived using the system throughput expression in (2.37), assuming that the PBS is responsible for the entirety of the system's energy consumption. Accordingly, the energy efficiency of the system can be expressed as

$$\eta_{sys} = \frac{\tau_{sys}}{P_p}, \tag{2.38}$$

where τ_{sys} and P_p represent system throughput and transmit power at PBS, respectively.

2.3.4 Ergodic Rate Analysis

The ergodic rate, expressed as $\mathcal{E} = \omega \times \mathbb{E}\left[\log_2(1 + SINR)\right]$, quantifies the long-term average data rate achievable under variable channel conditions, where ω indicates the bandwidth provisioned for the typical user [120, 130]. Serving as a fundamental indicator of spectral efficiency, a higher ergodic rate corresponds to improved channel capacity and superior communication performance. As outlined in Chap. 1, Sect. 1.8, this section systematically derives the ergodic rate expressions for various configurations, including the MBS tier, the PBS tier with and without NOMA, and the PBS scenario incorporating NOMA under i-SIC and non-linear HPA distortions.

Ergodic Rate of MBS Tier and PBS Tier (without NOMA)
The ergodic rate for a typical MU's is given by the following expression.

Proposition 2.3 *The ergodic rate of a typical MU is given by,*

$$\mathcal{E}_m = \frac{\omega_m}{ln2}\int_0^{\infty}\frac{1}{x}\left(1 - \mathcal{L}_{|\tilde{h}_{m_o}|^2}(xs_m)\right)\prod_t \mathcal{L}_{\Upsilon_t}(x\rho_t^I)e^{-x}dx, \tag{2.39}$$

where $s_m = \rho_m P_m$ denotes the effective received power at the typical MU, and ω_m represents the portion of the total bandwidth allocated to that user. The Laplace transform (LT) of the interference originating from the tth tier is denoted by $\mathcal{L}_{\Upsilon_t}(s)$, as defined in (2.24). Similarly, $\mathcal{L}_{|\tilde{h}_{m_o}|^2}(s)$ denotes the LT of the unordered channel gain for the MBS tier, which is evaluated as follows:

$$\mathcal{L}_{|\tilde{h}_{m_o}|^2}(s) = -\pi\lambda_m \mathcal{Y}_m^2 \sum_{\iota=0}^{I}\frac{b_\iota^m c_\iota^m}{c_\iota^m + s}. \tag{2.40}$$

The ergodic rate for the PBS tier under the OMA configuration can be derived analogously. It is formulated as follows:

$$\mathcal{E}_p = \frac{\omega_p}{ln2}\int_0^{\infty}\frac{1}{x}\left(1 - \mathcal{L}_{|\tilde{h}_{p_o}|^2}(xs_p)\right)\prod_t \mathcal{L}_{\Upsilon_t}(x\rho_t^I)e^{-x}dx, \tag{2.41}$$

where $s_p = \rho_p P_p$ represents the product of the transmit SNR and power at the PBS, while ω_p denotes the fraction of total bandwidth assigned to a typical PU. The Laplace Transform (LT) of the unordered channel gain in the PBS tier, denoted by $\mathcal{L}_{|\tilde{h}_{p_o}|^2}(s)$, is evaluated as follows:

$$\mathcal{L}_{|\tilde{h}_{p_o}|^2}(s) = -\sum_{\iota=0}^{I}\frac{b_\iota^p c_\iota^p}{c_\iota^p + s}. \tag{2.42}$$

Proof A detailed explanation is available in Appendix A.3. □

Ergodic Rate of PBS with NOMA, i-SIC & NL-HPA
This section focuses on analyzing the average achievable rate for the kth PBS user under NOMA with and without i-SIC and considering non-linear HPA characteristics.

Proposition 2.4 *For a PBS employing NOMA with perfect SIC and linear transmission, the ergodic rate at the kth user is expressed as:*

$$\mathcal{E}_k^p = \frac{1}{ln2}\int_0^{\infty}\frac{1}{x}\left(1-\mathcal{L}_{|h_k^{po}|^2}(xs_k^p)\right)\prod_{l=k+1}^{M_p}\mathcal{L}_{|h_k^{po}|^2}(xs_l^p)\times\prod_t \mathcal{L}_{\Upsilon_t}(x\rho_t^I)e^{-x}\,dx, \tag{2.43}$$

where $s_k^p=\rho_p P_p a_k$, $s_l^p=\rho_p P_p a_l$, *and* $\mathcal{L}_{|h_k^{po}|^2}(s)$ *is the LT of the PBS tier's ordered channel gain and is calculated as*

$$\mathcal{L}_{|h_k^{po}|^2}(s) = -\epsilon_k\sum_{r=0}^{M_p-k} f_1 f_2\times\sum_{T_k^r} f_3\left(\prod_{\iota=0}^{I}(b_\iota^p)^{q_\iota}\right)\frac{w_p}{s+w_p}, \tag{2.44}$$

where $w_p=\sum_{\iota=0}^{I} q_\iota c_\iota^p$, $f_1=\binom{M_p-k}{r}$, $f_2=\frac{(-1)^r}{k+r}$, *and* $f_3=\binom{k+r}{q_0+\ldots+q_N}$.

Proof A detailed explanation is available in Appendix A.4. □

Proposition 2.5 *Likewise, the ergodic rate for the kth user served by a PBS under NOMA with i-SIC and non-linear HPA is given by:*

$$\mathcal{E}_{k,iSIC}^{hpa,p} = \frac{1}{ln2}\int_0^{\infty}\frac{1}{x}\left(1-\mathcal{L}_{|h_k^{po}|^2}(xs_{k,iSIC}^{hpa,p})\right)\times\prod_{\substack{l=1\\ l\neq k}}^{M_p}\mathcal{L}_{|h_k^{po}|^2}(xs_{l,iSIC}^{hpa,p})\prod_t \mathcal{L}_{\Upsilon_t}(x\rho_t^I)e^{-x}\,dx, \tag{2.45}$$

where $s_{l,iSIC}^{hpa,p}=\rho_p P_p\left[a_l\beta_l\left(|\alpha_1|^2+|\alpha_3|^2 P_p^2\left|\sum_{i=1}^{M_p}\sqrt{a_i}\right|^4\right)+|\alpha_3|^2 P_p^2 a_k\left|\sum_{i=1}^{M_p}\sqrt{a_i}\right|^4\right]$, *and* $s_{k,iSIC}^{hpa,p}=\rho_p P_p a_k|\alpha_1|^2$, *and* $\mathcal{L}_{|h_k^{po}|^2}(s)$ *is the LT of ordered channel gain of PBS tier given in (2.44).*

Proof A detailed explanation is available in Appendix A.5. □

2.4 Analysis of Bounds

To attain better NOMA rates than OMA, we derive bounds on the power allocation factor in this section.

In our analysis, we enforce that each user's achievable rate under NOMA must be no less than its rate under OMA, i.e., ($Rate^{NOMA}\geq Rate^{OMA}$). This requirement can be written as:

$$\log_2(1+\gamma_{j\to k}^{p,NOMA}) \geq \frac{1}{2}\log_2(1+\gamma_p^{OMA}). \tag{2.46}$$

Specializing (2.46) to the two-user case—where a CEU with weaker channel and a CCU with stronger channel are jointly served by one BS—leads to the following bound on the power allocation coefficients:

$$a_{CEU} > \frac{\sqrt{1+\gamma}}{\gamma}(\sqrt{1+\gamma}-1) \triangleq a_{CEU}^*, \tag{2.47}$$

and

$$a_{CCU} < \frac{1}{\gamma}(\sqrt{1+\gamma}-1) \triangleq a_{CCU}^*, \tag{2.48}$$

equation (2.48) shows that if a_{CCU} surpasses the threshold a_{CCU}^*, the CCU's NOMA rate becomes inferior to its OMA rate. Furthermore, a_{CCU}^* is a monotonically decreasing function of the SINR threshold γ (for $\gamma > 0$), with its maximum value capped at 0.5, ensuring $0 < a_{CCU}^* < 0.5$.

Proposition 2.6 *Considering the i-SIC and applying the bounds on CCU,*

$$a_{CCU} > \frac{(1+\beta\gamma)(\sqrt{1+\gamma}-1)}{\gamma\{1+\beta(\sqrt{1+\gamma}-1)\}}. \tag{2.49}$$

Equation (2.49) establishes the minimum power allocation required for a given β. When this threshold is satisfied, the strong user's NOMA rate surpasses its OMA rate.

Proof A detailed explanation is available in Appendix A.6. □

Equations (2.48) and (2.49) illustrate the inherent trade-off between the strong and weak users' rates as a function of the power allocation coefficient. Thus, careful selection of power coefficients is important to ensure that both the CCU and CEU achieve higher individual rates under NOMA compared to OMA. The OMA baseline considered in this work assumes equal partitioning of time/frequency resources and uniform power distribution between the two users.

2.5 Results and Discussions

In Sect. 2.3, the outage probability expressions are analytically derived for MU and for both cell-center and cell-edge users in the PBS tier, incorporating the effects of power amplifier non-linearity under both NOMA and OMA transmission schemes. This section also provides a comprehensive performance evaluation in terms of outage probability, system throughput, and energy efficiency. To validate the accuracy

Table 2.1 Simulation Parameters

Parameters	Symbols	Value
Transmitting power from MBS, PBS	P_m, P_p	40 W, 1 W
MBS, PBS, User's density	λ_m, λ_p, λ_u	5×10^{-5} m^{-2}, 10^{-4} m^{-2}, 5×10^{-4} m^{-2}
Power allocation coefficients for NOMA	a_k	0.8, 0.2
Coverage range of MBS, PBS tier, respectively	$\mathcal{Y}_m$, $\mathcal{Y}_p$	1000 m, 5 m
Path loss exponent for MBS, PBS	ν_m, ν_p	3, 4
Non-linearities coefficients	α_1, α_3	1, 0.15
Target data rate (CEU, CCU)	R_1, R_2	0.1 bps, 0.05 bps
Residual interference parameter due to i-SIC	β	0.1
Number of NOMA users	M_p	2
Transmission bandwidth of the entire system	–	1 Hz

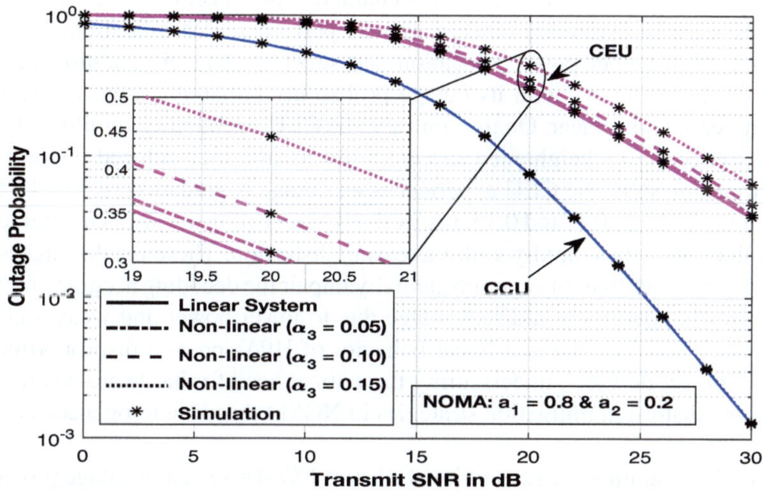

Fig. 2.2 Outage in linear versus Non-Linear NOMA systems with varying non-linearity

of the derived closed-form expressions, extensive simulations are conducted. The simulation settings, summarized in Table 2.1, are aligned with those adopted in prior studies such as [110, 112, 114, 125, 131, 138, 141].

Figure 2.2 illustrates the variation in outage probability for CEU and CCU as a function of transmit SNR under different non-linearity levels, denoted by parameter

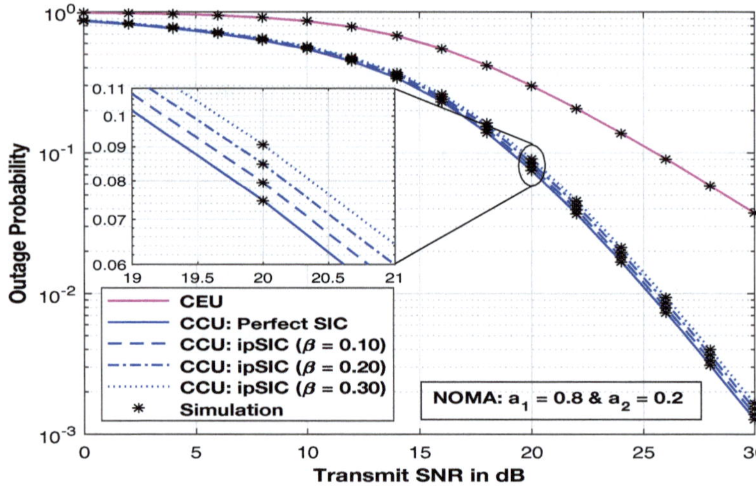

Fig. 2.3 Outage of linear NOMA system under varying residual interference (β) at CCU

α_3. The system operates under a NOMA transmission framework, with power allocation coefficients and non-linear amplifier characteristics specified in the figure. As α_3 increases, a stark asymmetry emerges in user performance degradation. The CCU, benefiting from higher power allocation and superior channel conditions, demonstrates strong resilience, with its outage probability remaining largely stable even under increased non-linear. On the contrary, the CEU's reliability significantly deteriorates, reflecting its heightened susceptibility to amplifier-induced distortion. At 20 dB transmit SNR, the CEU experiences an outage degradation of 3.8, 17.1, and 48.5% for $\alpha_3 = 0.05, 0.1$, and 0.15, respectively—revealing a non-linear degradation pattern that intensifies rapidly with increasing distortion. These results suggest that even moderate non-linearity can substantially impair the decoding accuracy for CEU, which already operates at a disadvantage due to lower power and weaker channel conditions. Therefore, the non-linear behavior of HPAs poses a disproportionately negative impact on CEU, underscoring the critical need for hardware-aware power control and distortion mitigation strategies in NOMA-based heterogeneous deployments.

Figure 2.3 examines the sensitivity of a linear NOMA system's outage probability to varying levels of residual interference (β) arising from i-SIC. The figure includes the power allocation settings for both CEU and CCU users. In this setup, the CCU decodes its message after performing SIC, whereas the CEU directly decodes its signal while treating the CCU's signal as noise. Under i-SIC, however, the CEU's high-SNR signal is not completely eliminated at the CCU, leading to partial interference remaining in the composite signal. This residual component corrupts the CCU's decoding process, thereby elevating its outage probability. Conversely, the CEU remains unaffected by i-SIC, as it does not rely on interference cancellation. Quantitatively, the CCU experiences a notable rise in outage probability—by 6.45, 13.53,

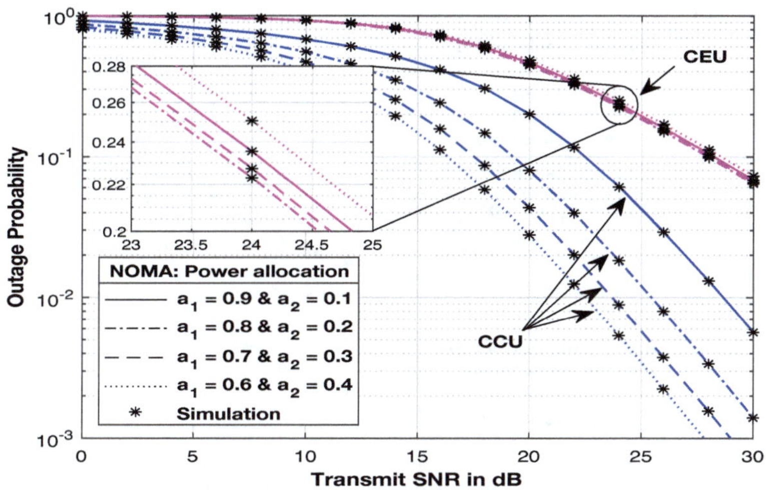

Fig. 2.4 Performance Comparison of Non-Linear NOMA ($\alpha_1 = 1$, $\alpha_3 = 0.15$) with i-SIC for Varying Power Ratios

and 21.33% at residual interference levels of $\beta = 0.1$, 0.2, and 0.3, respectively—when operating at 20 dB transmit SNR.

A broader comparison of Figs. 2.2 and 2.3 reveals distinct degradation patterns introduced by non-linearity and i-SIC. While the former exposes the vulnerability of the CEU to hardware-induced non-linear distortion, the latter highlights the adverse impact of residual interference on the CCU. This contrast underscores a critical insight: although both impairments are detrimental, non-linearity exerts a more profound influence on the overall system performance. Given that NOMA aims to boost spectral efficiency primarily by enhancing CEU reliability, the presence of non-linearities directly contradicts this objective. Therefore, from a system design perspective, addressing non-linear distortions should take precedence over mitigating i-SIC effects, as the former poses a more fundamental threat to the efficacy of NOMA in HCN deployments.

Figure 2.4 explores the impact of varying power allocation coefficients on the outage performance of the proposed non-linear NOMA system under imperfect SIC conditions. As the CCU's power allocation factor (a_2) is increased to 0.2, 0.3, and 0.4, a substantial enhancement in system performance is observed—yielding respective improvements of 59.9, 78.3, and 86.1% compared to the reference case of $a_2 = 0.1$, at a fixed transmit SNR of 20 dB. These results underscore the CCU's sensitivity to increased power allocation, which directly translates into improved signal reliability under non-linear distortion and residual interference. Since the CEU's power allocation coefficient ($a_1 = 1 - a_2$) inversely varies with that of the CCU, a trade-off naturally emerges. Initially, the CEU benefits slightly from reduced mutual interference as a_2 increases; however, as its allocated power diminishes further, the performance gain is reversed, leading to degradation in outage behavior. The findings

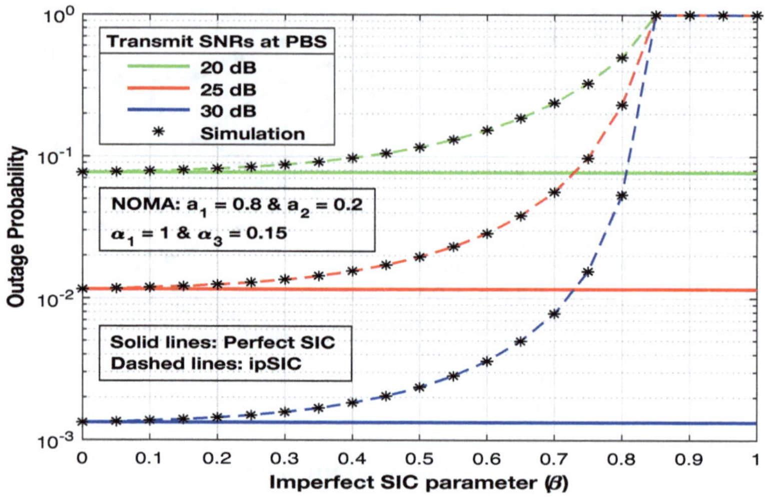

Fig. 2.5 Outage versus β in non-linear NOMA at CCU

reflect a clear asymmetry in how power allocation influences user performance: while the CCU consistently benefits from higher a_2, the CEU's performance peaks only within a narrow range before deteriorating—highlighting the importance of carefully optimized power control in non-linear NOMA deployments with imperfect SIC.

Figure 2.5 illustrates the variation in outage probability of the proposed non-linear system with respect to the residual interference factor (β) for transmit SNR levels of 20, 25, and 30 dB, under both ideal and imperfect SIC conditions. A clear degradation in performance is observed as β increases, highlighting the system's sensitivity to residual interference. At $\beta = 0.85$, the outage probability deteriorates significantly—by factors of 12.95 and 86.28—at 20 dB and 25 dB transmit SNRs, respectively. Beyond this threshold $\beta > 0.85$, the outage probability saturates at unity, indicating complete performance breakdown. This saturation occurs due to the substantial residual interference from the high-SNR CEU signal, which overwhelms the decoding process of the weaker CCU signal, effectively rendering i-SIC ineffective under severe non-linear distortion.

Figure 2.6 presents a comparison between NOMA and OMA in a an HCN, considering i-SIC and the impact of non-linear distortions introduced by HPAs. As depicted, users in the PBS tier—namely the CCU and CEU—demonstrate notably better performance than the MU. The advantages of NOMA are particularly evident within the PBS, where both CCU and CEU outperform their OMA counterparts. Under non-linear conditions, the outage probability for CCU and CEU improves by 28.88 and 52.10%, respectively, at a transmit SNR of 20 dB when employing NOMA with i-SIC compared to OMA. This gain stems from more balanced power distribution under NOMA. In contrast, under OMA, the base station allocates minimal power to the CEU, resulting in degraded performance; however, CEU performance is significantly enhanced in NOMA due to more favorable power allocation.

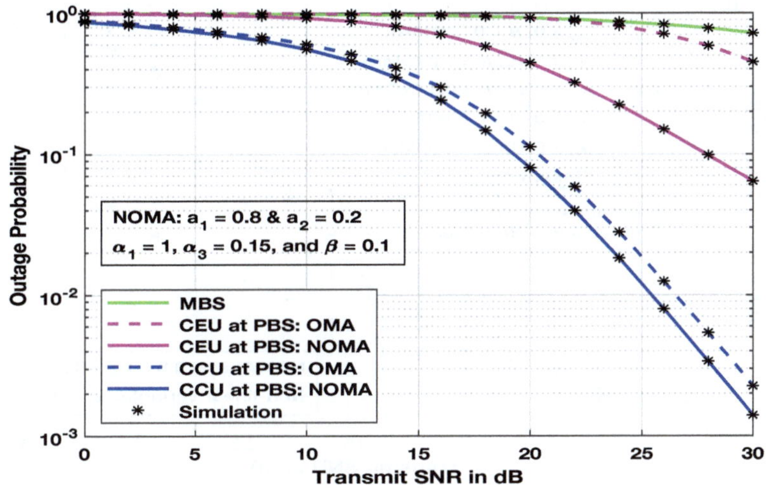

Fig. 2.6 Outage comparison of OMA and NOMA under i-SIC in non-linear systems

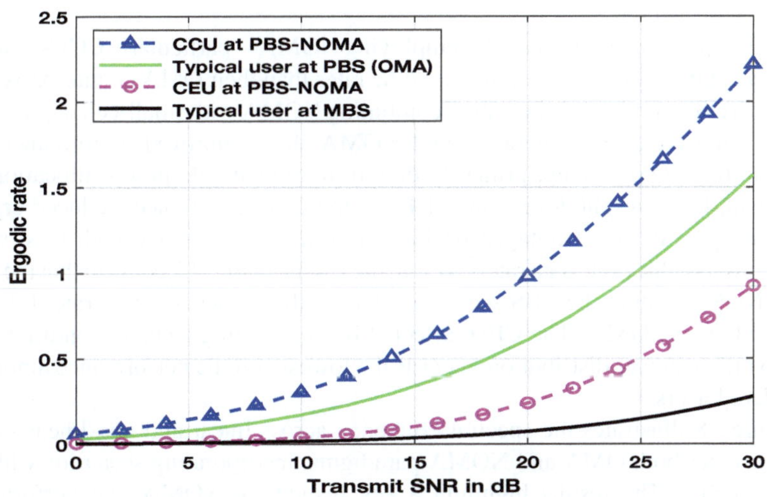

Fig. 2.7 Ergodic rate comparison of MU/PU in OMA versus NOMA

Figure 2.7 presents a comparison of ergodic rates for various user types—namely, MU and PU—under OMA and NOMA schemes, incorporating i-SIC and non-linear HPA effects. The figure clearly indicates that users served by the PBS consistently outperform those in the MBS in terms of ergodic rate. This performance gain is primarily due to the advantages inherent in picocell deployment, such as reduced interference levels and better link reliability. In contrast, MUs in the MBS suffer from lower ergodic rates, largely as a result of elevated inter-cell interference and limited resource allocation caused by the broader coverage area of macrocells. A more

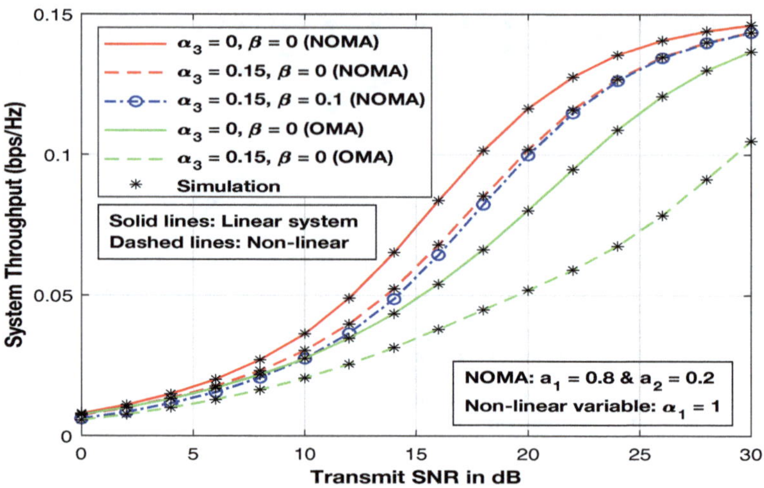

Fig. 2.8 Throughput of OMA versus NOMA with i-SIC in linear and non-linear systems

detailed inspection reveals that PUs employing NOMA—particularly CCUs—within the PBS exhibit significantly higher ergodic rates than their OMA counterparts. This underscores the synergistic benefit of combining NOMA with small-cell deployment. On the other hand, PUs operating under OMA show improved performance over CEUs in the NOMA setup, primarily due to more favorable power allocation and lower path loss. Despite leveraging NOMA, the CEU experiences reduced ergodic rates owing to its disadvantaged position in terms of SINR and path loss, which constrains its achievable data rate. Overall, the results in Fig. 2.7 confirm that the CCU under NOMA within the PBS scenario achieves the highest ergodic rate, followed by the PU using OMA. The CEU, affected by challenging channel conditions and less favorable power distribution, records the lowest ergodic performance among all considered users.

Figure 2.8 illustrates the spectral efficiency across linear and non-linear system models under both OMA and NOMA paradigms, incorporating scenarios with and without i-SIC. The results highlight a key insight: in NOMA, the performance enhancement at the CCU—achieved by allocating more signal power—more than compensates for the degradation at the CEU due to reduced power allocation. This strategic power imbalance leads to a net gain in system throughput, enabling NOMA to surpass OMA even in the presence of i-SIC. Additionally, the analysis reveals a pronounced throughput degradation in non-linear systems compared to linear ones across both access schemes. Interestingly, despite the throughput reduction caused by i-SIC in NOMA at the CCU, non-linear NOMA still delivers superior spectral efficiency at high SNRs when compared to both linear and non-linear OMA configurations, underscoring its potential in high-power regimes.

Figure 2.9 provides a comprehensive illustration of the energy efficiency performance for NOMA and OMA schemes in the context of non-linear systems, incor-

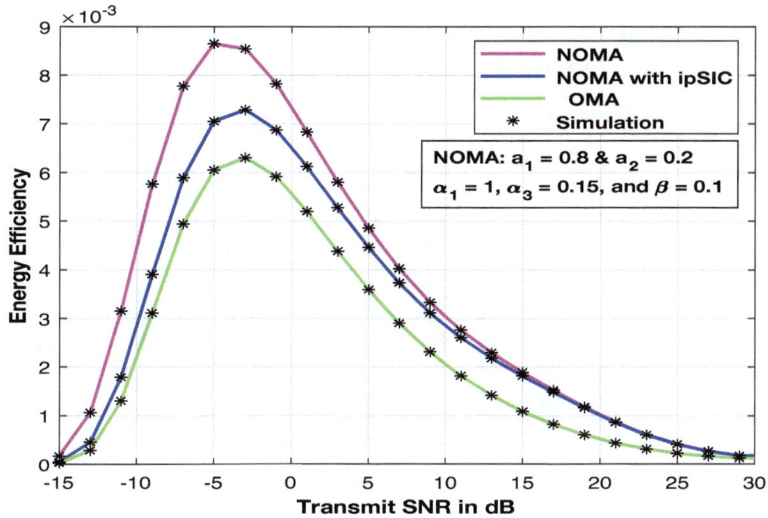

Fig. 2.9 Energy efficiency of non-linear OMA versus NOMA with i-SIC

porating both ideal and imperfect SIC scenarios. The analysis highlights a clear advantage of NOMA over OMA across the entire SNR spectrum, even when the effects of i-SIC are present. Notably, the configuration employing perfect SIC in NOMA achieves the highest energy efficiency, underscoring the critical role of accurate interference cancellation. In contrast, while NOMA with i-SIC still maintains a performance edge over OMA, its efficiency is notably diminished due to residual interference. These results underscore the sensitivity of NOMA's energy efficiency to SIC accuracy, particularly in the presence of hardware-induced non-linearities.

Figure 2.10 presents a quantitative comparison of system throughput for both linear and non-linear systems, with and without the presence of i-SIC. The analysis highlights the throughput gains achieved by employing NOMA with perfect SIC over OMA, showing improvements of 53, 45, 20, and 6% in linear systems, and 73, 96, 81 and 36% in non-linear systems at transmit SNR levels of 15, 20, 25, and 30 dB, respectively. The bar graph consistently illustrates the superior performance of NOMA over OMA across the entire SNR range. Furthermore, it emphasizes that NOMA yields a more pronounced performance gain margin in non-linear systems compared to linear counterparts. However, the introduction of i-SIC at the CCU adversely affects outage performance, leading to a noticeable degradation in both system throughput and energy efficiency, as depicted in Figs. 2.8 and 2.9, respectively. The performance advantage that NOMA offers over OMA in both linear and non-linear systems is significantly diminished in the presence of i-SIC. Specifically, when NOMA with i-SIC is employed, the observed performance gains in linear systems are 51, 45, 20, and 7%, and in non-linear systems, 63, 93, 80, and 37%, at transmit

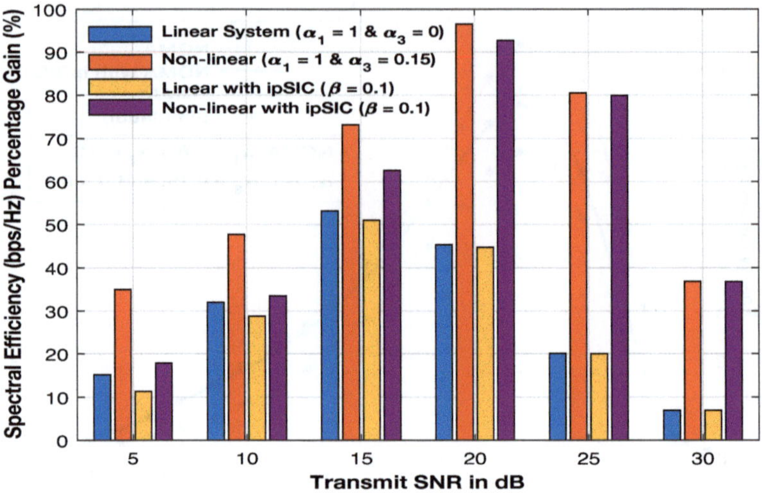

Fig. 2.10 Spectral efficiency of linear/non-linear systems with/without i-SIC: NOMA versus OMA (% Gain)

SNRs of 15, 20, 25, and 30 dB, respectively. These gains are notably lower than those achieved under ideal SIC conditions, underscoring the detrimental impact of i-SIC on overall system performance.

2.6 Summary

In this chapter, we examine the effects of non-linear HPA distortion on downlink NOMA performance—considering both ideal SIC and i-SIC at the CCU—and benchmark these results against an OMA reference. Our analysis demonstrates that HPA non-linearities and i-SIC significantly degrade energy efficiency, system throughput, and outage performance in HCN-NOMA deployments, particularly under high data-rate requirements. Numerical evaluations further reveal that, at specific transmit SNR levels and when the HPA operates in its non-linear region, NOMA can nearly double system performance relative to OMA. Although both linear and non-linear configurations employing i-SIC experience incremental performance losses, the non-linear system with i-SIC nonetheless exhibits a larger gain margin over its linear counterpart when using NOMA.

Chapter 3
Performance Analysis of NOMA-HCN with SIC and CSI Imperfections

Building upon the analysis presented in Chap. 2, which explores the impact of HPA non-linearity and i-SIC on NOMA-enabled HCNs, this chapter extends the investigation by incorporating the additional challenge of i-CSI. While Chap. 2 highlights the performance gains of NOMA over OMA despite non-linear distortions, this chapter further examines how imperfections in both SIC and CSI collectively influence key performance metrics such as outage probability, system throughput, and energy efficiency. The insights gained from Chap. 2 regarding the resilience of NOMA in the presence of non-linear distortions serve as a foundation for the analysis in this chapter, where we assess the combined effects of i-SIC and i-CSI on spectral and energy efficiencies in multi-tier HCNs. By integrating the concepts of imperfect SIC and CSI, this chapter provides a more comprehensive understanding of the practical limitations of NOMA, offering a comparative study with OMA under more realistic network conditions. Ultimately, the findings from this chapter, when combined with those from Chap. 2, contribute to a holistic perspective on the robustness of NOMA in HCNs, paving the way for designing efficient interference management and resource allocation strategies in next-generation wireless networks.

The rapid evolution toward 6G networks has intensified interest in ultra-dense networks (UDNs) leveraging small cells to overcome persistent challenges in coverage, latency, data throughput, and energy demands [106]. Among multiple access techniques, NOMA has emerged as a promising enabler, allowing simultaneous transmission to users with distinct power levels—particularly beneficial in scenarios where orthogonal resource allocation becomes infeasible [107]. In the context of cooperative NOMA, full-duplex (FD) user relaying demonstrates superior performance over half-duplex (HD) configurations under low-SNR conditions [144]. To support high user densities and escalating data demands, NOMA has been integrated into low-power PBS, forming NOMA-enabled HCNs. UAV-assisted HCN architectures that combine NOMA-based aerial links with mmWave terrestrial base stations have shown enhanced performance in terms of outage probability and ergodic sum-rate, enabled by optimal power allocation and random user distributions [109]. In the

V. Bhatia et al., *Advanced NOMA Techniques for Heterogeneous Cellular Networks*, Wireless Networks, https://doi.org/10.1007/978-3-032-09551-0_3

domain of wireless power transfer, power beacon-assisted Co-NOMA has been evaluated using non-linear energy harvesting models [145], while other studies explored SWIPT-based NOMA systems with spatially random users [112, 140], albeit under ideal assumptions of perfect SIC and CSI. Further investigations addressed NOMA with imperfect SIC (i-SIC) in multi-relay energy harvesting setups [123, 127], and examined the influence of non-linear power amplifier distortions assuming perfect CSI [146]. A hybrid NOMA-OMA architecture has also been proposed to enhance spectral efficiency by managing power and channel overlaps under simultaneous i-SIC and i-CSI constraints [147]. In contrast to prior studies, this work uniquely explores the integration of NOMA within HCNs under realistic conditions involving both i-SIC and i-CSI, modeled via stochastic geometry—an area that remains underexplored in current literature.

This chapter introduces several significant contributions, detailed as follows:

- The study conducts a comprehensive performance comparison between HCNs employing NOMA and those using OMA. It accounts for both ideal and imperfect SIC conditions. By deriving closed-form expressions for outage probability (OP) and its asymptotic behavior in the presence of i-CSI, the analysis offers a deeper understanding of system robustness under non-ideal conditions.
- To realistically capture the spatial randomness of BS deployments in HCNs, the PPP is employed due to its analytical tractability and reliability in modeling large-scale wireless networks. This approach enables the spatial characterization of both MBSs and PBSs, laying the foundation for performance analysis in randomly distributed network environments.
- Recognizing that precise CSI is essential for the effectiveness of SIC, the study addresses how i-CSI degrade decoding performance. An analytical framework is developed to evaluate the impact of i-CSI on users distributed according to a PPP in NOMA-enabled HCNs, incorporating both perfect and imperfect SIC scenarios. This framework helps quantify the performance loss due to realistic estimation errors.
- The interaction between i-SIC and i-CSI is systematically investigated to reveal their combined effects on network behavior. Rather than examining them in isolation, the study captures how their interplay influences system-level metrics such as spectral and energy efficiency. Detailed graphical illustrations are provided to support the analysis, offering visual insights that enrich the theoretical findings.

3.1 System Model

This work considers a two-tier HCN comprising MBSs and NOMA-enabled PBSs, leveraging NOMA to improve system adaptability in environments with diverse user channel conditions and QoS demands. As depicted in Fig. 3.1, both MBS and PBS deployments are modeled using PPP, offering a spatially grounded framework

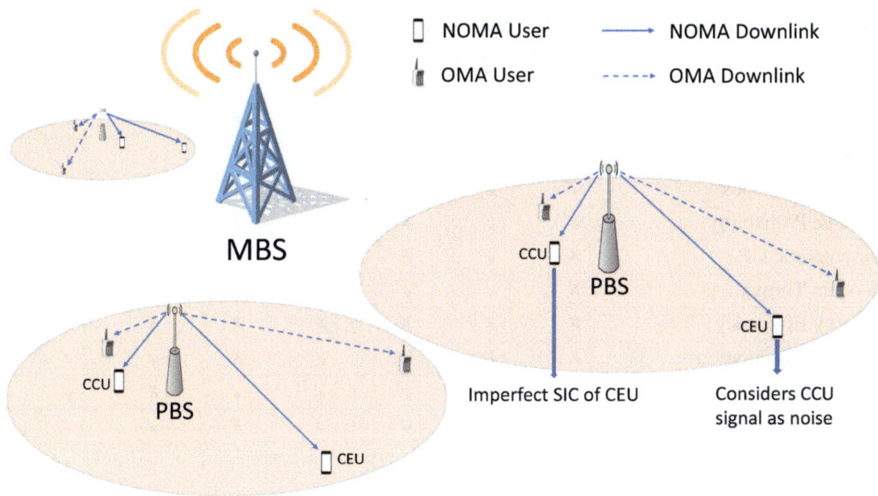

Fig. 3.1 System model for HCN-NOMA

for evaluating network performance and facilitating robust system-level analysis. Base stations across both tiers are independently distributed following PPPs Ω_t with densities λ_t for $t \in \{m, p\}$, while users are modeled using a separate PPP Ω_u with density λ_u. All nodes are assumed to operate in half-duplex mode and are equipped with a single antenna. Within the NOMA scheme, a_n denotes the power allocation factor for the nth user. The variables $\mathcal{Y}_t$ and P_t represent the coverage area and transmission power associated with tier t, respectively. User association with MBSs is governed by a nearest-neighbor policy. The signal attenuation is characterized using a bounded path loss model, expressed as $P(r) = \frac{1}{1+r_t^{\nu_t}}$ [145], where r_t is the user-to-BS distance and ν_t is the path loss exponent for the tth tier. The composite channel gain perceived by a user in tier t is defined by $|h_t|^2 = |\tilde{h}_t|^2 P(r)$, with $\tilde{h}_t$ drawn from a circularly symmetric complex Gaussian distribution $\mathcal{CN}(0, 1)$. The system-wide bandwidth is fixed 1 Hz, and R denotes the user-specific target data rate. Given that PBSs typically operate with more constrained backhaul capacities than MBSs, efficient CSI estimation becomes essential. Accordingly, the analysis herein concentrates on the influence of imperfect CSI at the PBS tier, where estimation and feedback overhead must be minimized to preserve system efficiency. CSI estimation is carried out using the minimum mean square error (MMSE) technique. Under this model, the actual channel coefficient $h_{t,i}$ for the ith user in tth tier is represented as the sum of the estimated component $\hat{h}_{t,i}$ and an error term e_i, i.e., $h_{t,i} = \hat{h}_{t,i} + e_i$, with $e_i \sim \mathcal{CN}(0, \sigma_e^2)$ as per [127]. Hence, the estimate $\hat{h}_{t,i}$ follows as a zero-mean complex Gaussian distribution with variance $\left(1 + r_{t,i}^{\nu_t}\right)^{-1} - \sigma_e^2$. The actual channel coefficients $\tilde{h}_t$ follow a standard complex Gaussian distribution. Channel gains are further categorized into ordered and unordered sets, denoted by $|\hat{h}|^2$ and $|\hat{\hat{h}}|^2$, respectively. Ordered gains are sorted based on received signal strength, aiding

Table 3.1 Comparison of the presented work with other related works

	[114]	[140]	[147]	[127]	[148]	[149]	This work
HCN multi-tier analysis	✓	✗	✗	✗	✗	✗	✓
NOMA	✓	✓	✓	✓	✓	✓	✓
PPP distributions	✓	✓	✗	✗	✗	✗	✓
Outage Probability	✓	✓	✗	✓	✓	✓	✓
Asymptotic OP	✗	✓	✗	✓	✓	✓	✓
System Throughput	✗	✓	✓	✓	✓	✗	✓
Energy Efficiency	✗	✗	✗	✗	✗	✗	✓
i-CSI Consideration	✗	✗	✓	✓	✗	✓	✓
i-SIC analysis	✗	✗	✓	✓	✓	✗	✓
Joint Impact Analysis	✗	✗	✓	✓	✗	✓	✓

in decisions such as user scheduling and resource allocation. Conversely, unordered gains preserve the native sequence of estimates without any form of prioritization (Table 3.1).

3.1.1 SINR at the MU Served by the MBS

The mathematical representation of the signal transmitted by the MBS is formulated to characterize the composite downlink transmission in a multi-user heterogeneous cellular network. Specifically, the transmitted signal expression serves as the foundation for subsequent analysis of received signal quality, interference levels, and achievable system performance under both ideal and non-ideal system conditions.

$$X_{m,tx} = x_m \sqrt{P_m}. \tag{3.1}$$

The signal model incorporates the principles of NOMA, where multiple user signals are superimposed in the power domain. The signal received at the typical MU is expressed as

$$X_{m,rx} = \underbrace{x_m \sqrt{P_m} \ddot{h}_{m_o}}_{\text{Intended desired signal}} + \underbrace{\sum_{t\in(m,p)} x_t \sqrt{P_t} \sum_{r\in\Omega_t\setminus\{m_o\}} \ddot{h}_r}_{\text{Intra-tier and Inter-tier interference by BSs}} + n_m, \tag{3.2}$$

where x_m denotes the intended signal for the MU,$\ddot{h}_i$ represents the unordered channel coefficient between the ith BS and the MU, where $i \in \{m_o, r\}$, and n_m corresponds to the additive channel noise. Accordingly, the SINR at the selected MU is given by [146].

$$\gamma_m = \frac{\rho_m P_m |\ddot{h}_{m_o}|^2}{\sum_t \rho_t^\iota \mathbb{I}_t + 1}, \tag{3.3}$$

where the transmit SNR for the MBS tier is defined as $\rho_m = \mathbb{E}\left[x_m^2\right]/\sigma_m^2$, where σ_m^2 denotes the noise variance. Similarly, the transmit SNR of interfering BSs is represented by $\rho_t^\iota = P_t/\sigma_m^2$, and the corresponding interference from the tth tier is given by $\rho_t^\iota \mathbb{I}_t$. Based on Slivnyak's theorem [133, 134], and without loss of generality, the selected MU is assumed to be located at the origin, with its serving MBS located at m_o. Under this setting, intra-tier interference from the MBS tier at the selected MU is modeled as $\rho_m^\iota \mathbb{I}_m$, where $\mathbb{I}_m = \sum_{i \in \Omega_m \setminus \{m_o\}} |\ddot{h}_i|^2$, and $|\ddot{h}_i|^2$ denotes the overall channel gain from the ith MBS to the MU. Consequently, $|\ddot{h}_{m_o}|^2$ represents the total unordered channel gain from the designated MBS to the selected MU. Likewise, the inter-tier interference from the PBS tier is denoted as $\rho_p^\iota \mathbb{I}_p$, where $\mathbb{I}_p = \sum_{j \in \Omega_p} |\ddot{h}_j|^2$, and $|\ddot{h}_j|^2$ denotes the total channel gain from the jth PBS transmitter to the MU.

3.1.2 SINR at the PBS User with Imperfect CSI (OMA Case)

The signal transmitted by the PBS is mathematically expressed as $X_{p,tx} = x_p \sqrt{P_p}$, while the corresponding signal received at the PBS-associated user (PU) is given by

$$X_{p,rx} = x_p \sqrt{P_p}\left(\hat{\ddot{h}}_{p_o} + e_p\right) + \underbrace{\sum_{t \in (m,p)} x_t \sqrt{P_t} \sum_{r \in \Omega_t \setminus \{p_o\}} \ddot{h}_r}_{\text{Intra-tier and Inter-tier interference by BSs}} + n_p, \tag{3.4}$$

where x_p denotes the intended signal for the PU, $\hat{\ddot{h}}_j$ represents the estimated unordered channel coefficient between the jth BS and the PU, where $j \in \{p_o, r\}$, and n_p corresponds to the additive noise. Accordingly, the SINR at the selected PU is expressed as [146].

$$\gamma_p = \frac{\rho_p P_p |\hat{\ddot{h}}_{p_o}|^2}{\sum_t \rho_t^\iota \mathbb{I}_t + \rho_p \sigma_e^2 + 1}, \tag{3.5}$$

where the transmit SNR at the PBS tier is defined as $\rho_p = \mathbb{E}\left[x_p^2\right]/\sigma_p^2$, where σ_p^2 denotes the noise variance. The transmit SNR from the interfering tth tier is given by $\rho_t^\iota = P_t/\sigma_p^2$, which accounts for the interference at the selected PU from other network tiers. Without loss of generality, and based on Slivnyak's theorem [146], the selected PU is assumed to be located at the origin, while the serving PBS is positioned at p_o. Under this assumption, the intra-tier interference from the PBS tier at the PU is represented as $\rho_p^\iota \mathbb{I}_p$, where $\mathbb{I}_p = \sum_{j \in \Omega_p \setminus \{p_o\}} |\ddot{h}_j|^2$, with $|\ddot{h}_j|^2$ denoting the total channel gain from the jth PBS to the PU. Similarly, the inter-tier interference from the MBS tier is expressed as $\rho_m^\iota \mathbb{I}_m$, where $\mathbb{I}_m = \sum_{i \in \Omega_m} |\ddot{h}_i|^2$, and $|\ddot{h}_i|^2$ represents the overall channel gain from the ith MBS to the selected PU.

3.1.3 SINR at PBS User (i-CSI, NOMA with I-SIC)

The PBS employs power-domain NOMA to serve M_p users, where the power allocation coefficients $a_1 \geq \cdots \geq a_{M_p}$ are assigned based on the ascending order of the users' channel gains, $|h_1^{p_o}|^2 \leq \cdots \leq |h_{M_p}^{p_o}|^2$. Within this framework, a 'near user' or 'CCU' is characterized by a stronger channel gain, while a 'far user' or CEU experiences a weaker channel gain. Assuming equal signal power across all NOMA users, the intended signal for the ith user is denoted by x_i, and the composite transmitted signal from the PBS is given by $X_{p,tx} = \sum_{i=1}^{M_p} x_i \sqrt{a_i P_p}$. The signal received at the kth user is expressed as

$$X_{k,rx}^{p} = \left(\hat{h}_k^{p_o} + e_k\right) s_k \sqrt{P_p} + \underbrace{\sum_{t \in (m,p)} x_t \sqrt{P_t} \sum_{r \in \Omega_t \setminus \{p_o\}} h_r}_{\text{Intra-tier and Inter-tier interference by BSs}} + n_k, \tag{3.6}$$

where s_k comprises of

$$s_k = \sqrt{a_k} x_k + \underbrace{\sum_{u=1}^{k-1} \sqrt{\beta_u} \sqrt{a_u} x_u}_{\text{i-SIC}} + \underbrace{\sum_{u=k+1}^{M_p} \sqrt{a_u} x_u}_{\text{Inter-User Interference}} . \tag{3.7}$$

In the downlink i-SIC scenario, users with stronger channel conditions may fail to fully decode the high SNR signal, leading to imperfect cancellation of the corresponding component in the NOMA transmission. As a result, these users experience residual interference from the undetected portion of the high SNR message during the decoding process. The kth user applies SIC by decoding the message intended for the jth user (with $j \leq k$); accordingly, the SINR at the kth user is expressed as

$$\gamma_{j \to k}^{p,i-SIC} = \frac{\rho_p P_p a_j \left|\hat{h}_k^{p_o}\right|^2}{\rho_p P_p \left|\hat{h}_k^{p_o}\right|^2 \sum_{\substack{n=1 \\ n \neq k}}^{M_p} \beta_n a_n + \sum_t \rho_t^l \mathbb{I}_t + \rho_p \sigma_e^2 + 1}, \tag{3.8}$$

where β_n equals 1 if $n > j$, and equals β otherwise, where a_n is the power-allocation coefficient for user $n \in \{k, j\}$. Under i-SIC, residual interference is modeled by setting $\beta_n = \beta; \forall \left\{n \in (1, 2, ..., M_p - 1)\right\}$, so that a fraction β of the high-SNR symbols remains in the superposed NOMA signal ($0 \leq \beta \leq 1$). Thus, $\beta = 0$ corresponds to ideal SIC (no residue), while $\beta = 1$ corresponds to no SIC. The user at the largest distance from the BS does not perform SIC and hence treats all other signals as noise. The SINR for the kth user decoding its own message is given by

$$\gamma_k^{p,i-SIC} = \frac{\rho_p P_p a_k \left|\hat{h}_k^{p_o}\right|^2}{\rho_p P_p \left|\hat{h}_k^{p_o}\right|^2 \sum_{\substack{n=1 \\ n \neq k}}^{M_p} \beta_n a_n + \sum_t \rho_t^{\iota} \mathbb{I}_t + \rho_p \sigma_e^2 + 1}, \tag{3.9}$$

where β_n is defined as 1 if $n > k$, and β otherwise.

3.2 Performance Analysis

This section evaluates the rate OP across the tiers defined in Sect. 3.1. In general, OP measures the probability that a wireless link experiences unacceptable performance degradation, such as diminished signal strength, reduced data rate, or elevated error rate.

3.2.1 *Outage Analysis for MBS Tier*

The outage probability for a user employing OMA and connected to the MBS is specified as:

$$\begin{aligned} \mathbb{P}_O^m &= \Pr\left[W_m \times \log_2(1+\gamma_m) < R\right], \\ &= \Pr\left[|\ddot{h}_{m_o}|^2 < \frac{\phi_m}{\rho_m P_m}\left(1 + \sum_t \rho_t^{\iota} \mathbb{I}_t\right)\right], \end{aligned} \tag{3.10}$$

where W_m denotes the bandwidth allocated to the OMA-based MU, γ_m represents the SINR at the MU as defined in (3.3), and R is the user's target data rate. Accordingly, the SINR threshold is given by $\phi = 2^{2R} - 1$. Assuming a homogeneous PPP and employing polar coordinates, the CDF of the unordered channel gain for the MBS tier is expressed as [146]:

$$F_{|\ddot{h}_{m_o}|^2}(y) = 2\pi\lambda_m \int_0^{\mathcal{Y}_m} (1 - e^{-(1+r_m^{\upsilon_m})y}) e^{-\pi\lambda_m r_m^2} r_m dr_m. \tag{3.11}$$

Using G-C quadrature [139], (3.11) can be approximated as

$$F_{|\ddot{h}_{m_o}|^2}(y) \simeq \pi\lambda_m \mathcal{Y}_m^2 \sum_{q=0}^{Q} b_q^m e^{-c_q^m y}. \tag{3.12}$$

Thus, the outage probability, as expressed in (3.10), can be reformulated as,

$$\mathbb{P}_O^m \approx \pi \lambda_m \mathcal{Y}_m^2 \sum_{q=0}^{Q} b_q^m e^{-c_q^m \frac{\phi}{\rho_m P_m}(1+\sum_t \rho_t^t \mathbb{I}_t)},$$
$$\approx \pi \lambda_m \mathcal{Y}_m^2 \sum_{q=0}^{Q} b_q^m e^{-c_q^m \frac{\phi}{\rho_m P_m}} \prod_t \mathcal{L}_{\mathbb{I}_t}\left(s_m \rho_t^t\right), \tag{3.13}$$

where Q strikes a balance between accuracy and complexity, $b_q^m = -w_Q\sqrt{1-\theta_q^2}\left(\frac{1}{2}\left(\theta_q+1\right)\right)e^{-\pi\lambda_m\left(\frac{1}{2}(\theta_q+1)\mathcal{Y}_m\right)^2}$, $w_Q = \frac{\pi}{Q}$, $\theta_q = \cos\left(\frac{2q-1}{2Q}\pi\right)$, $b_0 = -\sum_{q=1}^{Q} b_q^m$, $c_q^t = 1 + \left(\frac{\mathcal{Y}_t}{2}\theta_q + \frac{\mathcal{Y}_t}{2}\right)_t^{\nu}$, $c_0 = 0$, $s_m = \frac{c_q^m \phi}{\rho_m P_m}$ [146].

Laplace transform (LT) of interference from the tth tier is $\mathcal{L}_{\mathbb{I}_t}(s)$ expressed [145] as

$$\mathcal{L}_{\mathbb{I}_t}(s) = e^{\pi\lambda_t(s^{\mu_t}\Gamma(1-\mu_t,s)-s^{\mu_t}\Gamma(1-\mu_t))}, \tag{3.14}$$

where $\mu_t = 2/\nu_t$, $\Gamma(z) = \int_0^\infty x^{z-1}e^{-x}dx$ and $\Gamma(z,x) = \int_x^\infty t^{z-1}e^{-t}dt$.

3.2.2 *Outage Analysis for PBS Tier with I-CSI (Without NOMA)*

The rate OP at a typical PU is determined by considering a uniform distance from the PBS, is given by

$$\mathbb{P}_O^p = \Pr\left[W_p \times \log_2(1+\gamma_p) < R\right],$$
$$= \Pr\left[|\hat{h}_{p_o}|^2 < \frac{\phi_p}{\rho_p P_p}\left(1+\rho_p\sigma_e^2 + \sum_t \rho_t^t \mathbb{I}_t\right)\right], \tag{3.15}$$

where W_p denotes the bandwidth allocated to the OMA-based PU, γ_p is the SINR at the PU as defined in (3.5), and R represents the target data rate. Accordingly, the SINR threshold is given by $\phi = 2^{2R} - 1$. Assuming a homogeneous PPP and using polar coordinates, the CDF of the unordered channel gain in the PBS tier is expressed as [141]:

$$F_{|\hat{h}_{po}|^2}(y) = \frac{2}{\mathcal{Y}_p^2}\int_0^{\mathcal{Y}_p}(1-e^{-\left(\frac{1}{1+r_p^{\nu_p}}+\sigma_e^2\right)^{-1} y})r_p dr_p. \tag{3.16}$$

By applying the G-C quadrature [141] to (3.16), we get

$$F_{\left|\hat{h}_{po}\right|^2}(y) \simeq \sum_{q=0}^{Q} b_q^p e^{-\frac{c_q^p}{1+c_q^p\sigma_e^2}y}. \tag{3.17}$$

Thus, the OP, as expressed in (3.15), can be reformulated as,

$$\begin{aligned}\mathbb{P}_O^p &\approx \sum_{q=0}^{Q} b_q^p e^{-\frac{c_q^p}{1+c_q^p\sigma_e^2}\left\{\frac{\phi_p}{\rho_p P_p}\left(1+\rho_p\sigma_e^2+\sum_t \rho_t^t \mathbb{I}_t\right)\right\}}, \\ &\approx \sum_{q=0}^{Q} b_q^p e^{-\frac{c_q^p\phi_p}{\rho_p P_p\left(1+c_q^p\sigma_e^2\right)}\left(1+\rho_p\sigma_e^2\right)} \prod_t \mathcal{L}_{\mathbb{I}_t}(\rho_t^t s_p),\end{aligned} \tag{3.18}$$

where $b_q^p = -w_Q\sqrt{1-\theta_q^2}\left(\frac{1}{2}\left(\theta_q+1\right)\right)$, $b_0 = -\sum_{q=1}^{Q} b_q^p$, and $s_p = \frac{c_q^p\phi_p}{\rho_p P_p\left(1+c_q^p\sigma_e^2\right)}$.

3.2.3 Outage Analysis for PBS Tier (NOMA with I-SIC and I-CSI)

Given a fixed target rate R specified by the user's QoS demands, the OP is defined as follows:

$$\mathbb{P}_k^{O,p} = \Pr\left[\gamma_k^{p,i-SIC} < \phi_k, \gamma_{j\to k}^{p,i-SIC} < \phi_j\right], \tag{3.19}$$

where $\gamma_{j\to k}^{p,i-SIC}$ and $\gamma_k^{p,i-SIC}$ are given in (3.8), and (3.9) respectively. Given that self-message decoding in SIC affects the OP, the expression for the kth user is formulated as:

$$\mathbb{P}_k^{O,p} = \Pr\left[|\hat{h}_k^{p_o}|^2 < \frac{\psi_{max,i-S}^{i-C}\left(1+\rho_p\sigma_e^2+\sum_t \rho_t^t \mathbb{I}_t\right)}{\rho_p P_p}\right]. \tag{3.20}$$

This results in the OP as

$$\mathbb{P}_k^{O,p} = F_{\left|\hat{h}_k^{p_o}\right|^2}(y), \tag{3.21}$$

where $y = \frac{\psi_{max,i-S}^{i-C}(1+\rho_p\sigma_e^2+\sum_t \rho_t^t\mathbb{I}_t)}{\rho_p P_p}$ and $\psi_{max,i-S}^{i-C} = \max\left(\psi_{(1,i-S)}^{i-C}, \psi_{(2,i-S)}^{i-C}, \ldots, \psi_{(k,i-S)}^{i-C}\right)$. For a user with $k=1$, OP is simply calculated with $y = \frac{\psi_{j,i-S}^{i-C}(1+\rho_p\sigma_e^2+\sum_t \rho_t^t\mathbb{I}_t)}{\rho_p P_p}$, as it decodes its own message without performing SIC. $\psi_{j,i-S}^{i-C} = \frac{\phi_j}{a_j - \phi_j\sum_{l=1, l\neq j}^{M_P} a_l\beta_l}$, where ϕ_j is defined as $2^{R_j}-1$, with R_j representing the target data rate for the jth user, and the parameter for residual interference remains consistent as specified after (3.8).

Under i-CSI, the connection between ordered and unordered PBS-tier channel gains, denoted by $F_{\left|\hat{h}_{po}\right|^2}(y)$, can be expressed as [141]:

$$F_{\left|\hat{h}_k^{po}\right|^2}(y) = \epsilon_k \sum_{d=0}^{M_p-k} \binom{M_p-k}{d} \frac{(-1)^d}{k+d} \left(F_{\left|\hat{h}_{po}\right|^2}(y) \right)^{k+d}. \tag{3.22}$$

By substituting (3.17) into (3.22) and applying the multinomial theorem, the CDF of the ordered channel gains is derived as

$$F_{\left|\hat{h}_k^{po}\right|^2}(y) \simeq \epsilon_k \sum_{d=0}^{M_p-k} \binom{M_p-k}{d} \frac{(-1)^d}{k+d} \sum_{T_k^d} \binom{k+d}{l_0+\ldots+l_N} \prod_{q=0}^{Q} (b_q^p)^{l_q} e^{-\sum_{q=0}^{Q} l_q \frac{c_q^p}{1+c_q^p \sigma_e^2} y}. \tag{3.23}$$

Assuming uniform distance from the PBS and incorporating ordered channel gains, the rate outage probability for the kth selected PU is expressed as

$$\mathbb{P}_k^{O,p} \approx \epsilon_k \sum_{d=0}^{M_p-k} \binom{M_p-k}{d} \frac{(-1)^d}{k+d} \sum_{T_k^d} \binom{k+d}{l_0+\ldots+l_N} \prod_{q=0}^{Q} (b_q^p)^{l_q}$$
$$\times e^{-\sum_{q=0}^{Q} l_q \frac{c_q^p \psi_{\max,i-S}^{i-C}}{\rho_p P_p \left(1+c_q^p \sigma_e^2\right)} \left(1+\rho_p \sigma_e^2\right)} \prod_t \mathcal{L}_{\mathbb{I}_t}(\rho_t^l s_p), \tag{3.24}$$

where $\epsilon_k = \frac{M_p!}{(k-1)!(M_p-k)!}$, $s_p = \frac{\psi_{\max,i-S}^{i-C} \sum_{q=0}^{Q} l_q \frac{c_q^p}{1+c_q^p \sigma_e^2}}{\rho_p P_p}$, $T_k^d = \left\{ (l_0, \ldots, l_N) | \sum_{i=0}^{N} l_i = k+d \right\}$, and $\binom{k+d}{l_0+\ldots+l_N} = \frac{(k+d)!}{l_0! \ldots l_N!}$.

3.2.4 Asymptotic Analysis

High SNR Approximation of MBS Tier

Asymptotic analysis in the high-SNR regime offers insights into the system's performance ceiling and its behavior in the absence of noise or interference. It also aids in identifying the dominant components within mathematical expressions as SNR increases. When $\rho_m \to \infty$, a high-SNR approximation of (3.13) can be obtained using the first-order expansion $e^{-x} \approx 1 - x$, valid for x, and is given by

$$\mathbb{P}_O^{m,\infty} \approx -\pi \lambda_m \mathcal{Y}_m^2 \sum_{q=0}^{Q} b_q^m \frac{c_q^m \phi}{\rho_m P_m} \prod_t \mathcal{L}_{\mathbb{I}_t}\left(s_m \rho_t^l\right). \tag{3.25}$$

High SNR Approximation of PBS Tier with i-CSI (Without NOMA)
High-SNR asymptotic analysis elucidates system behavior in the absence of noise and interference, emphasizing the impact of dominant terms in the mathematical formulation. As $\rho_p \to \infty$, a high-SNR approximation of (3.18) is obtained using the first-order expansion $e^{-x} \approx 1 - x$ for $x \approx 0$, and is given by

$$\mathbb{P}_{O}^{p,\infty} \approx -\sum_{q=0}^{Q} b_q^p \frac{c_q^p \phi_p \left(1 + \rho_p \sigma_e^2\right)}{\rho_p P_p \left(1 + c_q^p \sigma_e^2\right)} \prod_t \mathcal{L}_{\mathbb{I}_t}(\rho_t^l s_p). \tag{3.26}$$

BS Tier High-SNR Approximation (NOMA with i-SIC and i-CSI)
As $\rho_p \to \infty$, the high-SNR approximation of (3.24) can be derived using the first-order expansion $e^{-x} \approx 1 - x$ for $x \approx 0$, and is expressed as

$$\mathbb{P}_{k,\infty}^{O,p} \approx \epsilon_k \sum_{d=0}^{M_p-k} \binom{M_p-k}{d} \frac{(-1)^d}{k+d} \sum_{T_k^d} \binom{k+d}{l_0 + \ldots + l_N} \prod_{q=0}^{Q} \left(b_q^p\right)^{l_q}$$
$$\times \left(1 - \sum_{q=0}^{Q} l_q \frac{c_q^p \psi_{\max,i-S}^{i-C}}{\rho_p P_p \left(1 + c_q^p \sigma_e^2\right)} \left(1 + \rho_p \sigma_e^2\right)\right) \prod_t \mathcal{L}_{\mathbb{I}_t}(\rho_t^l s_p). \tag{3.27}$$

3.3 Research Findings and Analysis

Section 3.2 presents the analytical derivation of OP expressions for users associated with both the MBS and PBS tiers, considering the presence of i-CSI under OMA and NOMA frameworks, with and without i-SIC. This section provides a performance evaluation based on simulation results of OP, from which system throughput is inferred, followed by the computation of EE. Simulation-based validation is carried out to confirm the analytical findings. The key simulation parameters include the MBS transmit power $P_m = 40$ W, PBS transmit power $P_p = 1$ W, MBS node density $\lambda_m = 5 \times 10^{-5}$ m^{-2}, PBS node density $\lambda_p = 10^{-4}$ m^{-2}, and user node density $\lambda_u = 5 \times 10^{-4}$ m^{-2}. NOMA users are assigned power allocation coefficients $a_k = 0.8$ and 0.2. The coverage areas are defined as $\mathcal{Y}_m = 1000$ m for MBSs and $\mathcal{Y}_p = 5$ m for PBSs, with corresponding path loss exponents $\nu_m = 3$ and $\nu_p = 4$. The target data rates are set to $R_1 = 0.1$ bps for CEUs and $R_2 = 0.05$ bps for CCUs. The variance of the channel estimation error is fixed at $\sigma_e^2 = 0.005$, consistent with prior works such as [127, 141, 146].

Figure 3.2 illustrates the variation in OP for NOMA under different levels of i-CSI. As the SNR increases, the system exhibits noticeable improvement in outage performance, with the OP progressively decreasing. This improvement continues up to an SNR of approximately ≈35 dB, beyond which the system performance saturates, and an error floor becomes evident—highlighting the limitations imposed by i-CSI. The figure considers both CEUs and CCUs, whose respective power allocation

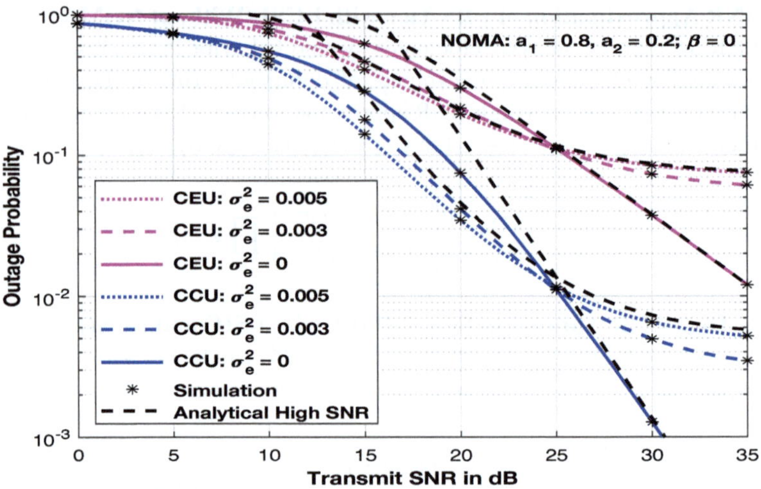

Fig. 3.2 Outage probability across multiple i-CSI parameter values utilizing NOMA

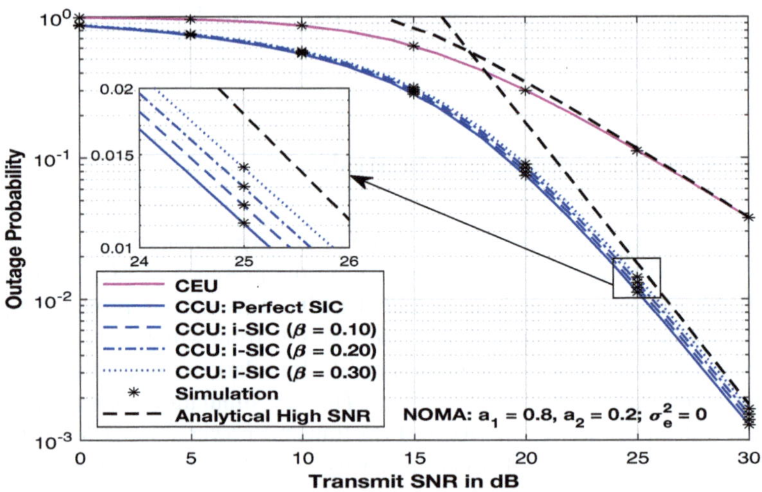

Fig. 3.3 Outage probability across multiple i-SIC parameter values utilizing NOMA at CCU

coefficients are clearly annotated within the plot. A key observation is the distinct impact of i-CSI on the outage behavior of CEUs and CCUs. Specifically, when the SNR is below or around 25 dB, increasing the i-CSI parameters results in improved OP for both user categories. However, at higher SNRs, the effect reverses: higher levels of i-CSI lead to a deterioration in outage performance. This trend emphasizes the dual nature of i-CSI's influence, benefiting low-SNR regimes while constraining performance in high-SNR scenarios.

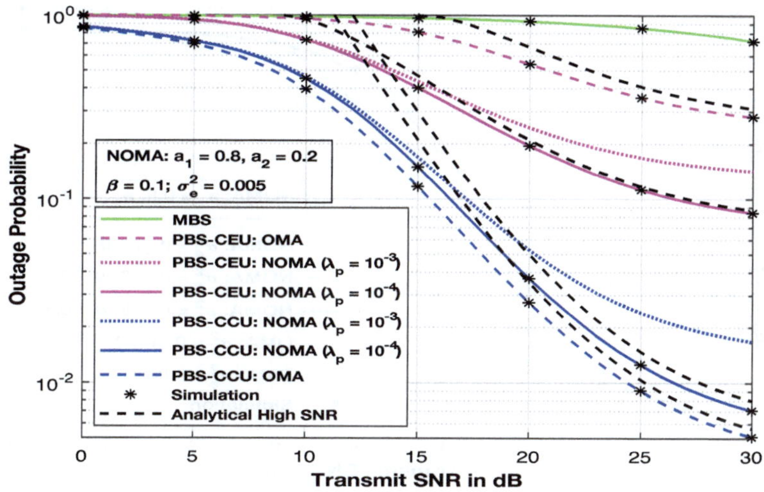

Fig. 3.4 Outage probability in the presence of i-SIC and i-CSI for both OMA and NOMA schemes

Figure 3.3 presents the OP performance of a NOMA system under different levels of residual interference resulting from i-SIC. As the SNR increases, the impact of i-SIC becomes more pronounced, particularly for the CCU. In high-SNR scenarios, the CCU experiences difficulty in accurately decoding and subtracting the signal of the CEU, whose transmission power is significantly higher. This leads to incomplete interference cancellation, leaving residual interference that degrades the CCU's performance. The figure clearly shows that the OP for the CCU is sensitive to the level of residual interference, higher i-SIC levels correspond to increased OP, indicating degraded reliability. In contrast, the CEU's performance remains largely unaffected across different i-SIC levels, as SIC operations are carried out solely at the CCU. This highlights the asymmetrical effect of i-SIC in NOMA systems, where only the user performing SIC (typically the CCU) experiences performance variation due to imperfect cancellation, while the CEU maintains stable performance irrespective of interference cancellation efficiency.

Figure 3.4 offers a comparative analysis of OMA and NOMA performance under the influence of both i-CSI and i-SIC within a HCN. The figure highlights that users associated with the PBS tier—namely, the CCU and CEU—achieve significantly better outage performance compared to MUs served by the MBS. Within the PBS tier itself, the CCU consistently outperforms the CEU across different access techniques, owing to its favorable channel conditions. The figure further reveals a nuanced trade-off in performance between OMA and NOMA. For the CEU, NOMA delivers superior outage performance due to its favorable power allocation in the superposition coding scheme. Conversely, the CCU benefits more from OMA, especially under the chosen power allocation coefficients, which prioritize CEU signal strength during NOMA decoding. A key insight from the plot is the impact of power allocation on CCU performance in NOMA. As the CCU's power allocation increases from

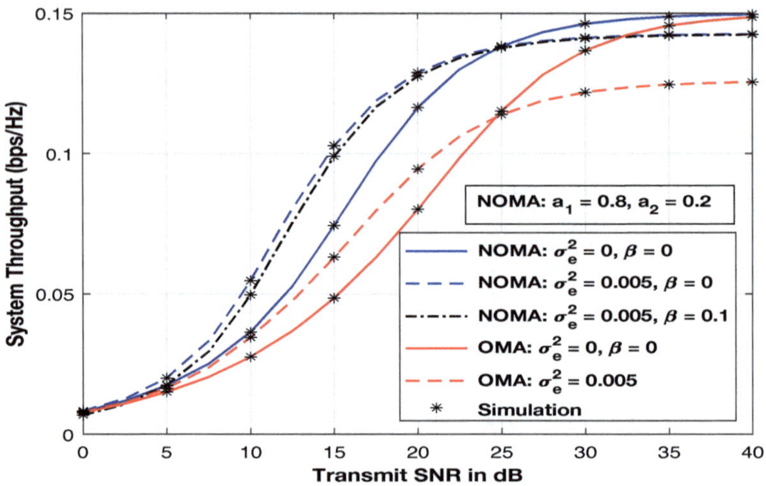

Fig. 3.5 Spectral efficiency in HCN using OMA and NOMA in the presence of i-SIC and i-CSI

0.1 to 0.2 and then to 0.3, the outage probability at the CCU improves significantly, achieving a performance gain of approximately 4.5 dB and 6.8 dB, respectively, at the OP threshold of 10^{-1}. This underscores the importance of optimal power distribution in enhancing system reliability in NOMA-based deployments. Furthermore, the effect of node density on user performance is clearly observed. A higher MBS node density improves MU performance by reducing the average user-to-BS distance, thus enhancing signal quality. Conversely, reducing the PBS node density benefits NOMA users by lowering intra-tier interference, thereby improving the reliability of user links in the PBS tier. Lastly, Figs. 3.2, 3.3, and 3.4 collectively validate the analytical derivations. A close match is observed between the simulated outage probability and the asymptotic expressions derived in (3.26) and (3.27), particularly in the high SNR regime. This agreement affirms the accuracy and practical relevance of the proposed mathematical models for performance evaluation in HCNs with NOMA.

The system throughput, denoted by τ_{sys}, is computed as $\tau_{sys} = \left(1 - \mathbb{P}_{CEU}^{O,p}\right) R_1 + \left(1 - \mathbb{P}_{CCU}^{O,p}\right) R_2$, where $\mathbb{P}_{CEU}^{O,p}$, and $\mathbb{P}_{CCU}^{O,p}$ represent the OP for the CEU and CCU, respectively, at the PBS tier. These outage probabilities are evaluated using the expression in (3.24) by setting $k = 1$ for CEU and $k = 2$ for CCU. The terms R_1 and R_2 correspond to the target data rates for the CEU and CCU, as defined in [145, 146]. Figure 3.5 presents the spectral efficiency performance of both OMA and NOMA systems under various scenarios, including the presence and absence of i-SIC and i-CSI. The results illustrate that an increase in signal power tends to improve the outage performance of the CCU significantly more than it degrades the performance of the CEU. This imbalance results in an overall enhancement in system throughput and spectral efficiency in NOMA compared to OMA. The application of i-CSI provides notable improvements in throughput, particularly in the low SNR regime

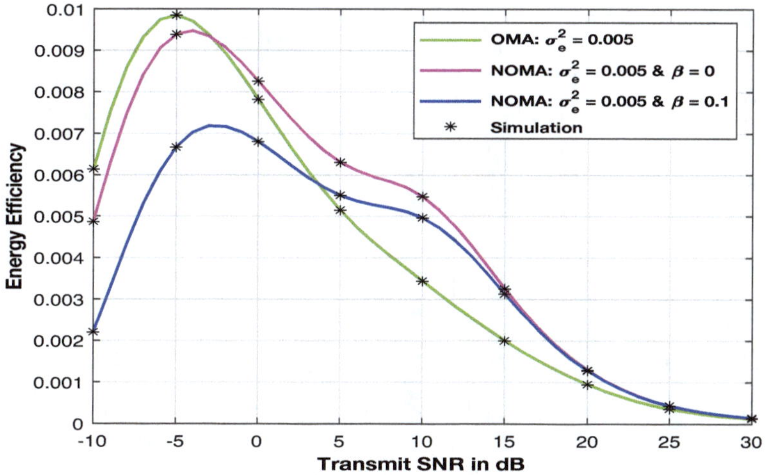

Fig. 3.6 Energy efficiency in HCN using OMA and NOMA in the presence of i-SIC and i-CSI

(below 25 dB). However, in high SNR scenarios, both NOMA and OMA experience performance degradation, though NOMA still maintains a throughput advantage. Despite the reduced throughput observed under i-SIC conditions, NOMA consistently demonstrates higher spectral efficiency than OMA across the entire SNR range. Furthermore, when perfect SIC and ideal CSI are assumed, NOMA distinctly outperforms OMA over all positive SNR values. Nevertheless, a clear degradation in system performance is evident when i-SIC is employed, highlighting the sensitivity of NOMA to SIC imperfections.

Energy efficiency (EE) of the system is quantified as the ratio of the total data transmitted successfully to the total power consumed, expressed as $\left(\eta_{sys} = \frac{\tau_{sys}}{P_p}\right)$, where τ_{sys} denotes the system throughput and P_p represents the transmit power at the PBS [146]. The metric provides a comprehensive evaluation of how efficiently the system utilizes power to deliver data. Figure 3.6 illustrates the EE performance for both OMA and NOMA systems, under scenarios with and without i-SIC and i-CSI. The results reveal that energy efficiency does not increase monotonically with SNR; instead, it reaches a peak at certain intermediate SNR values. Beyond this point, further increases in SNR lead to diminished EE, primarily due to the power consumption growing faster than the corresponding gain in throughput. At low SNR levels (i.e., SNR $\leq$25 dB), i-CSI significantly improves the outage performance, resulting in better EE. However, as SNR increases beyond this range, the benefits of i-CSI diminish, and EE performance begins to degrade. Importantly, the use of i-CSI within the moderate SNR range (5–25 dB) yields more substantial EE improvements compared to systems employing perfect CSI. This counter intuitive result stems from i-CSI's ability to balance system throughput and power consumption effectively in the presence of realistic channel imperfections. Moreover, NOMA systems leveraging perfect SIC and i-CSI consistently outperform OMA systems in terms of energy

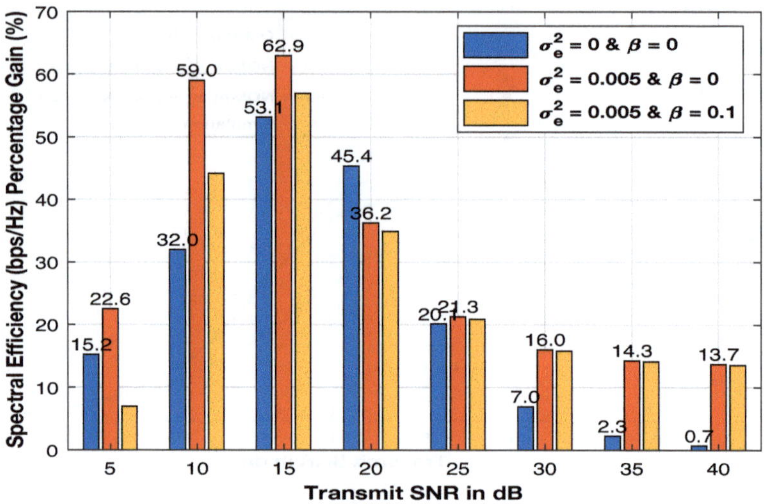

Fig. 3.7 Quantitative analysis: comparing NOMA versus OMA gain (%) with/without i-SIC and i-CSI

efficiency across all positive SNR regimes. However, the performance advantage of NOMA is notably reduced when i-SIC is employed, as imperfect interference cancellation adversely affects throughput while still incurring power costs, ultimately degrading the system's overall energy efficiency.

Figure 3.7 presents a quantitative comparison of system throughput for NOMA and OMA schemes under the influence of i-SIC and i-CSI. The analysis clearly demonstrates that NOMA consistently outperforms OMA across a range of transmit SNR values. Specifically, NOMA achieves throughput gains of approximately 57, 35, 21, and 16% over OMA at transmit SNR levels of 15, 20, 25, and 30 dB, respectively. These throughput enhancements, however, are not uniform across all SNR levels. The variation in performance gain can be attributed to the inherently non-linear behavior of the OP, as modeled in (3.24). Since system throughput is directly influenced by OP, the rate of improvement in NOMA's performance relative to OMA fluctuates with changing SNR conditions, leading to a non-monotonic trend in gain margins. Despite the performance degradation caused by i-SIC—particularly in the form of increased outage for the CCU—NOMA continues to maintain a consistent throughput advantage over OMA. Furthermore, the bar graph illustrates that the inclusion of i-CSI contributes more significantly to throughput gains than perfect CSI, especially in low-to-moderate SNR regimes. This observation underscores the potential of leveraging practical CSI estimation techniques to achieve improved system-level performance in NOMA deployments.

3.4 Summary

This chapter presents a comparative analysis of the effects of i-SIC and i-CSI on the performance of downlink NOMA and OMA systems. The results underscore the significant influence of these imperfections on key performance metrics—namely outage probability, system throughput, and energy efficiency—within the context of HCNs employing NOMA, particularly under high data rate demands. An increase in the density of MBS nodes is observed to enhance outage performance for macro-tier users, primarily due to reduced average distances between users and their serving MBSs, thereby improving signal quality. Conversely, a reduction in PBS node density proves beneficial for NOMA users by mitigating inter-cell interference, thus improving overall system performance in the proposed HCN-NOMA framework. Numerical evaluations further confirm that NOMA consistently outperforms OMA, especially in scenarios involving i-CSI. The advantage of NOMA becomes increasingly evident through its ability to maintain higher spectral efficiency across a broad range of SNR values, a benefit attributed to the inherent multiplexing capabilities of SIC. These findings strongly advocate for the adoption of NOMA in practical heterogeneous network deployments, where imperfections in interference cancellation and channel estimation are inevitable.

Chapter 4
SWIPT Enabled Cooperative NOMA in Heterogeneous Networks

Building upon the insights from Chap. 3, which examines the impact of i-SIC and i-CSI on the performance of NOMA-enabled HCNs, this chapter extends the discussion by incorporating SWIPT to improve the energy efficiency of cooperative NOMA systems. While Chap. 3 primarily focuses on the spectral and energy efficiency trade-offs in multi-tier HCNs under practical impairments, this chapter explores how SWIPT can be utilized to address the energy constraints faced by the CCU, which not only harvests energy from the received superimposed signal but also serves as a relay for the CEU. Since the CCU performs both energy harvesting and cooperative communication, its limited power supply poses a critical challenge, making efficient resource allocation essential for maintaining reliable transmission. The findings from Chap. 3, which highlight performance degradation due to imperfect CSI, provide a foundation for this chapter's investigation into the influence of imperfect CSI on SWIPT-based cooperative NOMA, where power allocation strategies must balance energy harvesting and relaying. Furthermore, this chapter expands on the interference analysis conducted in Chap. 3 by considering the role of aggregate interference in heterogeneous networks, particularly in the context of energy harvesting and cooperative transmission. By integrating concepts of interference management, cooperative NOMA, and SWIPT-enabled energy harvesting, this chapter presents a comprehensive framework for improving both spectral and energy efficiency while ensuring sustainable and reliable communication for power-constrained users.

Developing spectrally efficient and energy-sustainable solutions with low hardware complexity is essential for enabling practical and environmentally friendly beyond 5G wireless systems. Among various multiple access schemes, NOMA has emerged as a promising approach to support a massive number of users while enhancing spectral utilization [150]. NOMA enables simultaneous access for multiple users over the same orthogonal resource blocks, thereby optimizing spectrum use [151].

Simultaneous Wireless Information and Power Transfer (SWIPT) has also attracted significant research interest due to its potential in establishing energy-efficient communication infrastructures. In SWIPT, the receiver extracts both information and energy concurrently from the incoming RF signals. The idea was first pro-

V. Bhatia et al., *Advanced NOMA Techniques for Heterogeneous Cellular Networks*, Wireless Networks, https://doi.org/10.1007/978-3-032-09551-0_4

posed in [99] for point-to-point communication and later extended to scenarios with parallel channels in [152]. Two practical receiver models, namely time-switching (TS) and power-splitting (PS), were introduced for relay-assisted networks in [101]. SWIPT applications are particularly relevant in cooperative relaying networks, where decode-and-forward (DF) and amplify-and-forward (AF) protocols are commonly employed. For DF relaying, a simplified SWIPT protocol based on antenna switching was developed in [102]. In [104], the performance of SWIPT in DF-based cooperative systems with spatially distributed random relays was analyzed using stochastic geometry, considering multiple transmit nodes and a single destination. Regarding AF relaying, the TS and PS protocols introduced in [101] remain foundational. Further, resource allocation strategies have been studied to optimize SWIPT-enabled systems. For instance, [153] focused on designing efficient allocation algorithms for OFDMA-based SWIPT systems, while [154] addressed resource management in two-tier heterogeneous networks (HetNets) with NOMA and energy cooperation. The security of NOMA systems has also been investigated, such as in [155], which examined the secrecy performance of downlink NOMA in the presence of eavesdroppers, and [156], which introduced a secure cooperative NOMA transmission scheme using a two-slot strategy. Additionally, energy-efficient optimization of SWIPT-NOMA frameworks has been tackled in [157]. A robust resource allocation approach for cooperative NOMA HetNets under imperfect successive interference cancellation (SIC) conditions was explored in [158].

4.1 Motivation and Contribution

The current literature offers only a few studies addressing outage performance in SWIPT systems affected by multi-tier interference. For example, in [74], the authors analyze the outage probability (OP) and ergodic rate of a NOMA-based system with a single base station and demonstrate that NOMA outperforms OMA in terms of achievable sum rate. In [159], a cooperative NOMA framework integrated with SWIPT is proposed for a single base station (BS) scenario, where the outage performance under different user selection strategies is analyzed. Other earlier studies, such as [77], have also utilized cooperative NOMA concepts to enhance system performance. However, these works primarily focus on single BS setups and do not account for the complexities introduced by multi-tier heterogeneous networks (HetNets), including inter-tier interference, which can significantly impact the outage behavior of users.

Recent research has started to explore NOMA in the context of large-scale HetNets. For instance, [160] investigates outage probability in large-scale cognitive radio environments. Similarly, [161] examines user outage and achievable rate in a single-tier cellular network. The work in [56] considers a NOMA-enabled HetNet and evaluates key performance metrics such as coverage probability, ergodic rate, and energy efficiency. Moreover, [162] presents an analytical model for downlink NOMA in large-scale networks and analyzes its outage performance.

Despite these advancements, the integration of cooperative NOMA with SWIPT in the HetNet environment remains relatively unexplored. Specifically, the combined use of these techniques to enhance outage performance and system throughput in a HetNet setup—critical for enabling energy-efficient beyond-5G systems—has not been adequately addressed in [159–162]. Another important limitation in the existing literature is the prevalent assumption of perfect channel state information (CSI) at the BS. This assumption is unrealistic in real-world deployments due to the associated signaling overhead, channel estimation inaccuracies, and quantization errors, as noted in [163]. To address this gap, this chapter also investigates system performance under the practical constraint of imperfect CSI. The main contributions of the chapter are highlighted below.

- This work introduces an analytical model for a SWIPT-assisted cooperative NOMA scheme in a multi-tier heterogeneous downlink network. The goal is to enhance the performance of the CEU by leveraging assistance from the CCU. To ensure sustainable cooperation, CCUs harvest energy from received signals using SWIPT, thereby avoiding battery depletion during the relaying process. Unlike prior studies, the CCU in this model harvests energy from both the superimposed signal and the total co-tier and cross-tier interference—referred to as Energy Harvesting from Superimposed and Interfering Signals (EHSI). This combined harvesting approach has not been addressed in earlier research.
- In addition to incorporating EHSI, this work also explores the practical challenge of i-CSI and evaluates its impact on system reliability and performance.
- By employing stochastic geometry, the outage probabilities for both CCU and CEU are derived in closed form under the imperfect CSI condition and considering the EHSI model. These expressions are then used to analyze the system's overall throughput.
- A comparative analysis is conducted between scenarios with and without SWIPT at both CCU and CEU. Furthermore, the proposed EHSI-based model is benchmarked against conventional energy harvesting approaches where energy is extracted solely from the superimposed signal (EHS). The study offers fresh insights into the advantages of including interference in the energy harvesting process. The theoretical findings are corroborated through extensive Monte Carlo simulations.

4.2 System Model

The system under consideration features a two-tier heterogeneous wireless network composed of a MBS tier and a FBS tier. The spatial locations of the MBSs and FBSs are modeled as independent homogeneous PPPs, represented by Ω_t, where $t \in m, f$, with m and f referring to the MBS and FBS tiers, respectively, and their corresponding densities given by λ_t. Users are assumed to be uniformly distributed

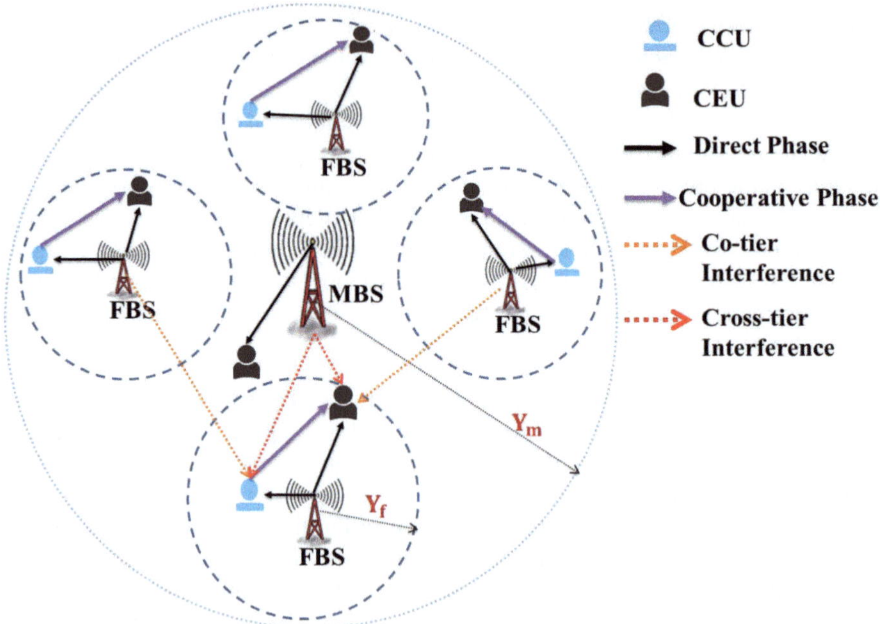

Fig. 4.1 Depiction of a downlink cooperative NOMA-SWIPT system model featuring a designated FBS and a NOMA user pair within a multi-tier network setup. The MBSs and FBSs are spatially distributed based on a PPP model

across the network with a user density of λ_u. The transmission powers of the MBS and FBS are denoted by P_m and P_f, respectively, while their coverage areas are represented by $\mathcal{Y}_m$ for the MBS tier and $mathcalY_f$ for the FBS tier. The performance evaluation is focused on a NOMA user pair within the FBS tier, where a CCU and a CEU are grouped based on their respective channel qualities. The system operates over a normalized transmission bandwidth 1 Hz, following the setup described in [114]. The total transmission time block is split into two equal-duration phases: a direct transmission phase and a cooperative relaying phase. During the direct transmission phase, the FBS sends a NOMA signal simultaneously to both the CCU and CEU. In the subsequent cooperative phase, the CCU acts as a DF relay with energy harvesting capability, forwarding the CEU's message after harvesting energy from the previous transmission, as illustrated in Fig. 4.1.

The wireless channel between the base station of the tth tier and a typical user i is modeled as $\frac{h_i}{\sqrt{1+r_i^{\alpha\_t}}}$, where $h_i \sim \mathcal{CN}(0, 1)$ represents small-scale fading and follows a circularly symmetric complex Gaussian distribution with zero mean and unit variance. The term r_i denotes the distance between the ith user and its serving BS, and α_t indicates the path loss exponent for tier t. Following the MMSE estimation error [164]:

$$\hat{h}_i = h_i + e_i, \tag{4.1}$$

where e_i denotes the channel estimation error associated with the ith user and follows a complex Gaussian distribution, i.e., $e_i \sim \mathcal{CN}(0, \sigma_e^2)$. The estimated channel coefficient for user i is represented by $\hat{h}_i$. Consequently, $\hat{h}_i$ is also modeled as a complex Gaussian random variable with zero mean and a variance of $(1 + r_{t,i}^{\alpha_t})^{-1} + \sigma_e^2$. The accuracy of the channel estimation is directly influenced by the value of σ_e^2.

4.2.1 SINR at CCU and CEU

Consider two users selected for NOMA transmission through a random user selection strategy, as suggested in [33]. The estimated channel gains of these users are arranged such that $\hat{h}_1 \leq \hat{h}_2$, while the corresponding power allocation coefficients satisfy $a_1 \geq a_2$. User-2, experiencing a stronger channel, is designated as the CCU, while User-1, having a weaker channel, is identified as the CEU. Accordingly, $\hat{h}_1$ represents the channel power gain for the CEU, and $\hat{h}_2$ denotes the channel power gain for the CCU.

Direct Transmission

In accordance with the NOMA transmission protocol, the femto base station (FBS) transmits a superimposed signal given by $X_f = \sum_{i=1}^{2} x_i \sqrt{a_i P_f}$ to serve its users, where x_i represents the signal designated for the ith user, with $\mathbb{E}[x_i^2] = 1$ for all $i \in 1, 2$. Each user receives the signal as:

$$Y_k = \hat{h}_k \sqrt{(1 + r_k^{\alpha})^{-1}} X_f + n_k, \tag{4.2}$$

where n_k denotes the additive white Gaussian noise at the kth user. At the CCU, successive interference cancellation (SIC) is applied, wherein the CCU first decodes the CEU's signal and then subtracts it to retrieve its own message. The SINR at the CCU for decoding the CEU's message is

$$\gamma_{2\to 1}^{1} = \frac{\beta \rho_f a_1 |\hat{h}_2|^2 (1 + r_2^{\alpha})^{-1}}{\beta \rho_f a_2 |\hat{h}_2|^2 (1 + r_2^{\alpha})^{-1} + \beta \sum_t \rho_t \mathcal{I}_t + \rho_f \sigma_e^2 + 1}, \tag{4.3}$$

where β denotes the power splitting ratio, and $\rho_t = \frac{P_t}{\sigma_f^2}$ represents the transmit SNR of tier t. The term $\rho_f \mathcal{I}_f$ accounts for co-tier interference from other FBSs and $\rho_m \mathcal{I}_m$ denotes cross-tier interference from MBSs. Following SIC, the SINR for the CCU to decode its own message becomes

$$\gamma_2^1 = \frac{\beta \rho a_2 |\hat{h}_2|^2 (1 + r_2^{\alpha})^{-1}}{\beta \sum_t \rho_t \mathcal{I}_t + \rho_f \sigma_e^2 + 1}. \tag{4.4}$$

It is assumed that the CCU is equipped with an energy-harvesting capability, as described in [101], and PS technique [159] to implement SWIPT. The received signal at the CCU is divided, with one portion allocated for information decoding and the other for energy harvesting to facilitate cooperative transmission. The energy harvested (E_2) at the CCU, sourced from both the intended transmission and the interference generated by surrounding MBSs and FBSs, is expressed as

$$E_2 = \frac{T}{2}\eta(1-\beta)\left(P_f|\hat{h}_2|^2(1+r_2^{\alpha})^{-1} + P_f\mathcal{I}_f + P_m\mathcal{I}_m\right), \tag{4.5}$$

where T is the total transmission time block, and η is the conversion efficiency of the energy harvester. The transmit power from CCU can be expressed as

$$\begin{aligned} P_2 &= \frac{E_2}{\frac{T}{2}} \\ &= \eta(1-\beta)\left(P_f|\hat{h}_2|^2(1+r_2^{\alpha})^{-1} + P_f\mathcal{I}_f + P_m\mathcal{I}_m\right), \end{aligned} \tag{4.6}$$

where P_2 denotes the harvested power from the energy harvester which is also the transmit power of the CCU. This chapter assumes that the energy required by the CCU for decoding information is negligible when compared to the energy used for data transmission during the cooperative communication phase. This simplification aligns with prior studies that incorporate power amplifier models, such as in [101, 165, 166]. The signal-to-interference-plus-noise ratio (SINR) at the CEU during the initial transmission phase is expressed as

$$\gamma_1^1 = \frac{\rho a_1|\hat{h}_1|^2(1+r_1^{\alpha})^{-1}}{\rho a_2|\hat{h}_1|^2(1+r_1^{\alpha})^{-1} + \sum_t \rho_t\mathcal{I}_t' + \rho_f\sigma_e^2 + 1}, \tag{4.7}$$

where $\rho_f\mathcal{I}'_f$ accounts for co-tier interference from neighboring FBSs to the CEU, and $\rho_m\mathcal{I}'_m$ represents cross-tier interference from MBSs to the CEU.

Cooperative Transmission

During the second phase of transmission, the CCU relays the information to the CEU. The corresponding SINR at the CEU for detecting the signal x_1 forwarded by the CCU is expressed as

$$\gamma_1^2 = \frac{P_2|\hat{h}_{12}|^2(1+r_{12}^{\alpha})^{-1}}{P_f\mathcal{I}'_f + P_m\mathcal{I}'_m + P_f\sigma_e^2 + \sigma_1^2}, \tag{4.8}$$

where $|\hat{h}_{12}|^2$ represents the channel between the CCU and CEU, and r_{12} is the separation between the two nodes. To enhance the received signal quality, the CEU applies selection combining (SC) to the transmissions received from both the FBS and CCU [167]. SC is favored over maximal ratio combining (MRC) in this context due to its lower implementation complexity at the receiver side [168]. Additionally, SC offers diversity performance similar to that of MRC, especially when the number of signal paths is less than five [169]. In this context, SC is implemented by selecting the higher SINR between the direct transmission phase (4.7) and the cooperative transmission phase (4.8). The resulting SINR at the CEU is given by

$$\gamma_1^{sc} = \max\left(\gamma_1^1, \gamma_1^2\right). \tag{4.9}$$

4.3 Performance Analysis

4.3.1 Outage Probability at CCU

The outage probability at the CCU is influenced by its ability to perform SIC on the CEU's signal. An outage event at the CCU can happen in two scenarios: (i) when it fails to decode the CEU's message, or (ii) when it successfully decodes the CEU's message but fails to decode its own. Thus, the total outage probability at the CCU is derived by considering both these possibilities. Therefore, the outage probability at CCU is calculated as

$$P_{out}^{cc} = \mathcal{P}\left(\gamma_{2\to1}^1 < \phi_1\right) + \mathcal{P}\left(\gamma_{2\to1}^1 \geq \phi_1, \gamma_2^1 < \phi_2\right), \tag{4.10}$$

where $\phi_1 = 2^{2R_1} - 1$ and $\phi_2 = 2^{2R_2} - 1$.

Proposition 4.1 *The OP of the CCU is given as*

$$P_{out}^{cc} = \frac{1}{2}\sum_{n=0}^{N} b_n e^{-\frac{c_n(\epsilon_1+\rho_f\sigma_e^2)}{1+c_n\sigma_e^2}} \mathcal{L}_{I_p}[s_p], \tag{4.11}$$

where $\mathcal{L}_{I_p}[s_p] = \mathcal{L}_{I_f}[s_f]\mathcal{L}_{I_m}[s_m]$, $s_f = \frac{\rho_f c_n \epsilon_1}{1+c_n\sigma_e^2}$, $s_m = \frac{\rho_m c_n \epsilon_1}{1+c_n\sigma_e^2}$, $\epsilon_1 = \frac{\phi_1}{\beta\rho(a_1-\phi_1 a_2)}$, $\epsilon_2 = \frac{\phi_2}{\beta\rho_f a_2}$, $b_n = -\omega_N(1+\theta_n)\sqrt{1-\theta_n^2}$, $b_0 = -\sum_1^N b_n$, $\omega_N = \pi/N$, $\theta_n = \cos\left(\frac{2n-1}{2N}\pi\right)$, $c_n = 1 + \left(\frac{y_f}{2}\theta_n + \frac{y_f}{2}\right)^\alpha$, $c_0 = 0$,

Proof Please refer to Appendix B.1. □

where N governs the trade-off between computational complexity and accuracy in the Gaussian-Chebyshev (G-C) quadrature approximation. In equation (4.11), the terms $\mathcal{L}_{I_f}(s)$ and $\mathcal{L}_{I_m}(s)$ represent the Laplace transforms (LT) of the interference from the FBSs and MBSs, respectively. The LT for the interference contributed by the tth tier is evaluated following the approach in [114] as

$$\mathcal{L}_{I_t}(s) = e^{\pi\lambda_t\left(s^{\delta_t}\Gamma(1-\delta_t,s) - s^{\delta_t}\Gamma(1-\delta_t)\right)}, \tag{4.12}$$

where $\delta_t = \frac{2}{\alpha_t}$, $\Gamma(a, x) = \int_x^\infty t^{a-1} e^{-t} dt$ and $\Gamma(z) = \int_0^\infty x^{z-1} e^{-x} dx$.

4.3.2 *Outage Probability at CEU*

Similarly, the CEU experiences an outage in two cases: first, if neither the CCU nor the CEU is able to decode the CEU's message; and second, if the CCU successfully decodes and forwards the message, but the overall received SINR at the CEU is insufficient to meet the required data rate. Therefore, the outage probability for the CEU is formulated as

$$P_{out}^{ce} = \underbrace{\mathcal{P}\left(\gamma_{2\to 1}^{1} < \phi_1, \gamma_1^{1} < \phi_1\right)}_{\Theta_1} + \underbrace{\mathcal{P}\left(\gamma_{2\to 1}^{1} > \phi_1, \gamma_1^{sc} < \phi_1\right)}_{\Theta_2}. \tag{4.13}$$

Proposition 4.2 *The OP at the CEU is given by*

$$P_{out}^{ce} = \frac{1}{4}\sum_{n=0}^{N}\sum_{k=0}^{K} b_n b_k e^{-\frac{(c_n+c_k)(\epsilon_1+\rho_f\sigma_e^2)}{(1+c_n\sigma_e^2)(1+c_k\sigma_e^2)}} \mathcal{L}_{I_p}[s_p]\mathcal{L}'_{I_p}[s_p] + \frac{1}{2}\sum_{n=0}^{N} b_n e^{-\frac{c_n(\epsilon_1+\rho_f\sigma_e^2)}{(1+c_n\sigma_e^2)}} \mathcal{L}_{I_p}[s_p] \times$$
$$\left(1 - \lim_{B\to\infty} \frac{\pi^2 B}{\alpha^2 N M \mathcal{Y}_f^3} \sum_{n=1}^{N} \left(1-\theta_n^2\right)^{\frac{1}{2}} \frac{e^{-K_n}}{K_n^{1+\frac{2}{\alpha}}} \left((2+\alpha K_n)\Gamma\left(\frac{2}{\alpha}\right) - \alpha K_n \Gamma\left(\frac{2}{\alpha}, \mathcal{Y}_f^{\alpha} K_n\right) \times \right.\right.$$
$$\left.\left. \sum_{m=1}^{M} b_m e^{-\frac{c_m Q_1}{(1+c_m\sigma_e^2)K_n}} - \alpha\Gamma\left(\frac{2+\alpha}{\alpha}, \mathcal{Y}_f^{\alpha} K_n\right)\right) \mathcal{L}_{I_p}\left[\frac{d_m Q \rho_m}{(1+d_m\sigma_e^2)K_n}\right]\right) \tag{4.14}$$

Proof Please refer to Appendix B.2. □

4.3.3 *System Throughput*

To assess the overall performance of the system, its throughput is analyzed. The system throughput is determined using the outage probabilities of both the CCU and CEU, assuming fixed target transmission rates, given as

$$T_{sys} = \left(1 - P_{out}^{ce}\right) R_1 + \left(1 - P_{out}^{cc}\right) R_2, \tag{4.15}$$

where P_{out}^{cc} and P_{out}^{ce} is obtained from (4.11) and (4.14).

4.3.4 *Proportional Fairness Index*

In NOMA systems, the total transmit power is divided among users with varying channel conditions, which can lead to unequal performance levels, where some users perform better than others [170]. As a result, ensuring user fairness is a critical aspect in NOMA. Ideally, the performance disparity between users should be minimized to maintain fairness. To evaluate this, the proportional fairness (PF) index is employed, which compares user performances based on their outage probabilities. If fairness is not preserved, one user may experience a significantly higher outage probability than the other. To address this imbalance, the PF index for outage probability has been introduced as a measure [171].

$$PF = \frac{P_{out}^{cc}}{P_{out}^{ce}}. \tag{4.16}$$

The expressions for P_{out}^{cc} and P_{out}^{ce} can be obtained from (4.11) and (4.14).

4.4 Results and Discussions

This section provides numerical evaluations based on the OP and system throughput expressions formulated in Sect. 4.4 for the proposed system architecture. The analytical findings are verified through Monte Carlo simulations with 10^5 iterations. The numerical parameters used include $\eta = 0.7$, $\beta = 0.5$, $N = 10$, $M = 10$, and $B = 1000$. Network parameters for both analytical and simulation results follow [114], with values $P_m = 40$ W, $P_f = 1$ W, $\lambda_m = 10^{-4}$ m^{-2}, $\lambda_f = 10^{-3}$ m^{-2}, $\mathcal{Y}_m = 1$ km, $\mathcal{Y}_f = 5$ m, $a_1 = 0.8$, $a_2 = 0.2$, $\alpha_f = 4$, and $\alpha_m = 3$. The channel estimation error variance σ_e^2 is considered at 0.005 and 0.008, while perfect CSI is modeled by setting $\sigma_e^2 = 0$. For the analysis, the transmit SNR at the FBS tier is varied from 0 to 40 dB, with comparisons conducted at 36 dB transmit SNR. In this chapter, EHSI-PPP denotes the proposed system where the CCU harvests energy from both the superimposed signal and aggregate co-tier and cross-tier interference, with BSs following a PPP distribution. EHS-SBS refers to the scenario where the CCU harvests energy only from the superimposed signal and the network includes a single FBS as in [159]. EHS-PPP indicates the CCU harvests energy only from the superimposed signal while BSs follow a PPP distribution. Systems labeled without EHSI-PPP, without EHS-PPP, and without EHS-SBS represent cooperative NOMA configurations without energy harvesting, as considered in [77].

Figure 4.2 illustrates the OP of the CCU as a function of transmit SNR. It is observed that the CCU with EHS-SBS experiences an 89.37% decline in outage performance compared to without EHS-SBS. This performance decline arises because the CCU equipped with EHS-SBS dedicates only part of its harvested energy to decoding information and utilizes just half of the time slot for cooperative communication. Similarly, the outage probability of the CCU with EHSI-PPP shows a 75.56% deterioration relative to the CCU without EHSI-PPP, due to the same underlying reason. When comparing the two, the CCU with EHS-SBS outperforms the CCU with EHSI-PPP by 25.52%, which can be attributed to the fact that the CCU with EHSI-

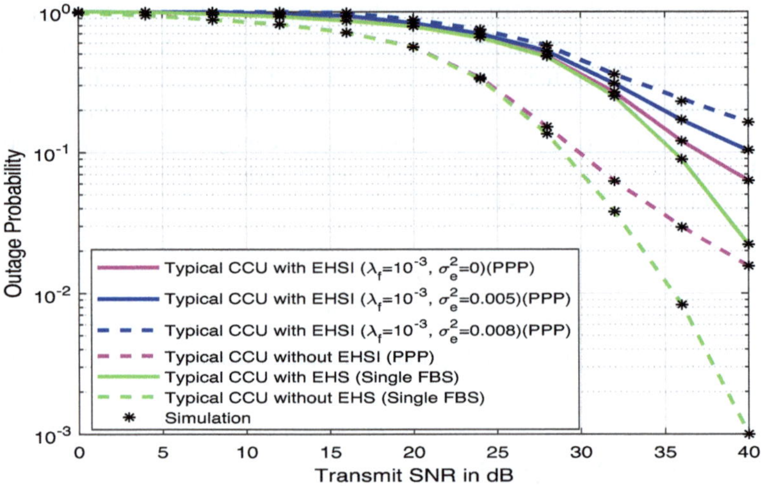

Fig. 4.2 Variation of outage probability at the CCU with transmit SNR

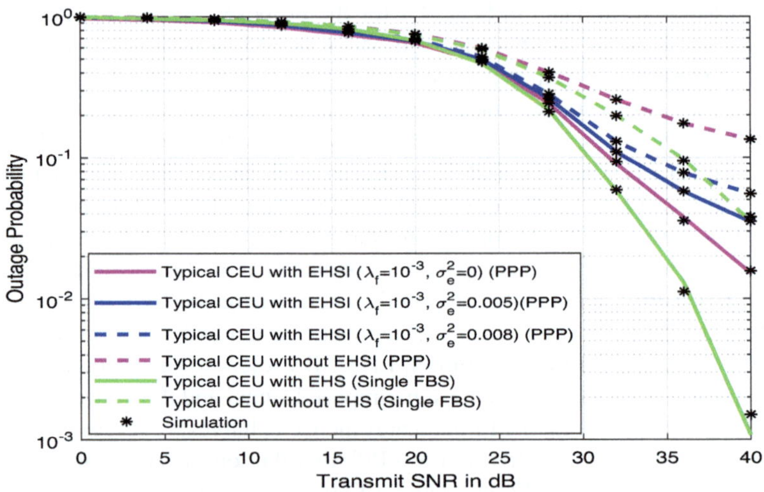

Fig. 4.3 Variation of outage probability at the CEU with transmit SNR

PPP is affected by both co-tier and cross-tier interference, whereas the CCU with EHS-SBS experiences no interference as the analysis assumes only a single BS.

Figure 4.3 presents the outage probability of the CEU as a function of transmit SNR. The results indicate that EHS-SBS achieves an 88.16% improvement in outage performance compared to the CEU without EHS-SBS. This improvement is attributed to the CEU using SC to detect signals from both transmission phases, whereas without EHS-SBS relies solely on the direct phase. Similarly, the CEU with EHSI-PPP shows a 79.63% enhancement over the CEU without EHSI-PPP for the

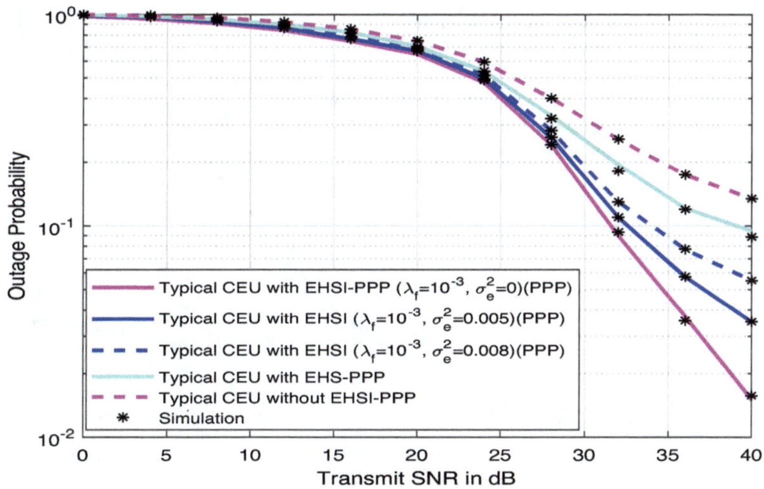

Fig. 4.4 Comparision of proposed EHSI-PPP system with EHS-PPP system

same reason. When comparing the two, EHS-SBS outperforms the EHSI-PPP by 68.45%. Additionally, Figs. 4.2 and 4.3 reveal that imperfect CSI results in worse outage probabilities compared to perfect CSI. This degradation is due to channel estimation errors, which act as interference. Although employing imperfect CSI reduces system complexity, it comes at the cost of higher outage probability. Moreover, increasing the error variance from 0.005 to 0.008 further degrades the outage performance, reflecting the impact of channel estimation quality.

Figure 4.4 illustrates that the CEU's outage performance with EHS-PPP improves by 31.49% compared to the system without EHS-PPP. Moreover, the proposed CEU employing EHSI-PPP demonstrates a 79.63% enhancement in outage performance relative to the system lacking EHSI-PPP. Additionally, the CEU with the proposed EHSI-PPP shows a 70.27% improvement over the CEU using EHS-PPP.

The overall system throughput is evaluated and presented in Fig. 4.5. It can be seen that at lower transmit SNR levels, the non-EH system outperforms EH-enabled systems in terms of throughput. This occurs because, at low SNR, the degradation in CCU's outage performance outweighs the gains achieved at the CEU. Additionally, since the harvested energy scales with transmit SNR, limited power is available for cooperation at low SNR values. However, at higher transmit SNR, the outage performance of the EH-enabled system improves and exceeds that of the non-EH system. Among the EH-based configurations, the proposed setup with EHSI-PPP at the CCU delivers superior throughput compared to the system with only EHS-PPP.

Figure 4.6 compares the user PF index across various systems: without EHS-SBS, without EHSI-PPP, with EHS-SBS, with EHS-PPP, and with EHSI-PPP, using the same parameters as those in the outage probability analysis with target rates $R_1 = R_2 = 0.5$. It is evident from Fig. 4.6 that the system employing EHSI-PPP achieves improved user fairness compared to all other setups, except the one with

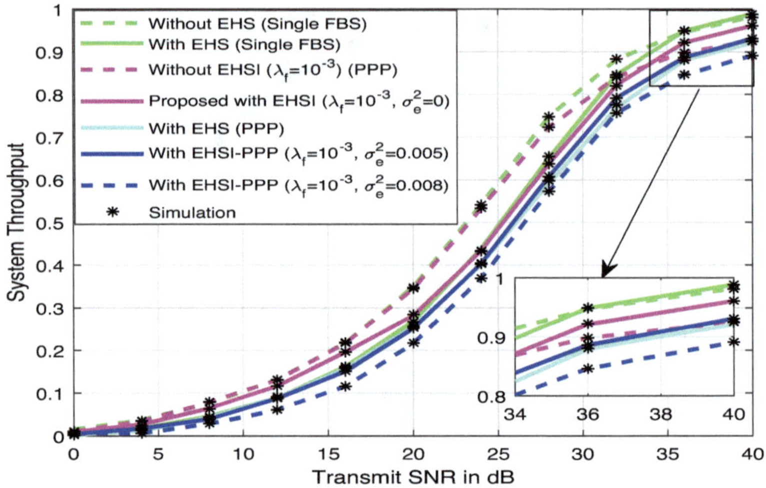

Fig. 4.5 System throughput versus transmit SNR for the target rates $R_1 = 0.5$ and $R_2 = 0.5$

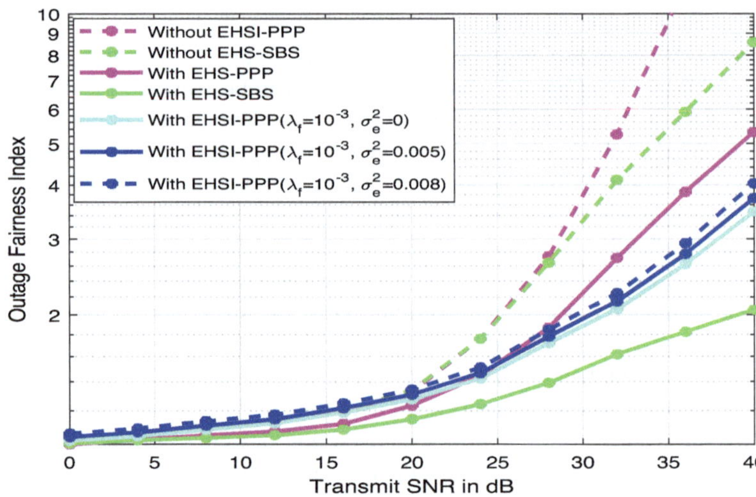

Fig. 4.6 Proportional fairness (PF) index versus transmit SNR

EHS-SBS. For systems with EHS-SBS, EHS-PPP, and EHSI-PPP, the CCU and CEU exhibit more balanced outage probabilities, resulting in a PF index below 2 up to a transmit SNR of 28 dB. On the other hand, systems without EHS-SBS and without EHSI-PPP show the greatest disparity in outage between users. This is because, in non-SWIPT configurations, the CEU relies solely on one transmission phase, whereas SWIPT-enabled systems allow the CEU to combine SINR from both phases.

4.5 Summary

This chapter investigates the implementation of SWIPT in a two-tier downlink cooperative NOMA network, where base stations are distributed according to a PPP and channel state information is imperfect. Analytical expressions for outage probability and system throughput are derived for both the CCU and CEU. The findings show that the CCU's outage performance deteriorates in the presence of energy harvesting signal interference modeled via PPP (EHSI-PPP). However, the CEU benefits significantly from the CCU's cooperation. The throughput analysis confirms that the proposed EHSI-PPP-assisted system surpasses conventional cooperative NOMA setups without EHSI-PPP at higher transmit SNR levels. Furthermore, PF index results highlight that the EHSI-PPP-based model ensures improved user fairness compared to other schemes, with the exception of the one utilizing EHS-SBS.

Chapter 5
Impulsive Noise Effects on NOMA Heterogeneous Networks

Expanding upon the investigation of SWIPT-enabled cooperative NOMA in heterogeneous networks from Chap. 4, this chapter explores the impact of impulsive noise (Im-N) on the performance of HCN-NOMA. While Chap. 4 examines how energy harvesting at the CCU facilitates cooperative communication for the CEU, it also highlights practical impairments such as imperfect CSI that affect system efficiency. These impairments, along with aggregate interference in heterogeneous networks, set the stage for these chapter's examination of the disruptive effects of impulsive electromagnetic interference, which challenges conventional Gaussian noise-based models. The study in this chapter extends prior analysis by considering how impulsive noise interacts with the stochastic spatial distribution of network nodes, influencing both spectral and energy efficiency. By integrating concepts of interference modeling, and advanced noise analysis, this chapter further refines the performance evaluation of NOMA-enabled HCNs, emphasizing its robustness under real-world noise conditions, particularly in smart home, grid, and industrial IoT environments.

As wireless networks evolve toward sixth-generation (6G) architectures, achieving ultra-reliable, high-capacity, and energy-efficient communication becomes increasingly critical. UDNs have emerged as a cornerstone in addressing these demands; however, the dynamic allocation of limited radio resources—especially under low-rate requirements or in degraded channel conditions—poses significant challenges. NOMA, by enabling multiple users to share the same spectrum resources via power-domain multiplexing, offers a compelling solution for efficient resource utilization in future UDNs [106, 107]. When integrated into HCNs, particularly with low-power PBSs, NOMA becomes a scalable mechanism to serve dense user populations with diverse QoS demands [114]. Recent advances extend NOMA's application to cooperative and wireless-powered HCNs, where SWIPT facilitates energy-efficient communication among randomly distributed users [172]. NOMA has also proven effective in aerial-terrestrial HCNs, where UAV-assisted mmWave deployments enhance outage and throughput performance through optimized power allocation strategies [109]. Moreover, research has examined the impact of hardware

V. Bhatia et al., *Advanced NOMA Techniques for Heterogeneous Cellular Networks*, Wireless Networks, https://doi.org/10.1007/978-3-032-09551-0_5

imperfections, such as non-linear distortions from high-power amplifiers, highlighting the practical limitations of NOMA with i-SIC in realistic deployments [146]. A coordinated direct and relay transmission (CDRT) protocol, incorporating i-SIC, has also been proposed for NOMA-based IoT networks with direct connectivity to IoT devices [128]. Hybrid networks leveraging multi-RAT capabilities—including mmWave and sub-6 GHz—have further demonstrated NOMA's potential in achieving massive connectivity [130]. Additionally, RIS-enhanced MIMO systems operating under the NOMA framework have outperformed OMA counterparts, with performance scaling favorably with the number of RIS elements [173]. Despite these advancements, most analytical studies assume idealized conditions dominated by AWGN, owing to its mathematical tractability under the central limit theorem. However, real-world environments—such as urban landscapes, high-voltage facilities, power lines, and industrial settings—are often dominated by Im-N arising from switching operations, electromagnetic interference, and partial discharges [174–177]. Despite their usefulness, earlier studies primarily focus on systems operating under AWGN. However, research indicates that systems designed for AWGN often suffer significant performance degradation in Im-N environments, and since power domain NOMA is particularly sensitive to such disturbances and variations in the amplitude, studying NOMA under fading and Im-N is essential for defining practical system limits and design. To address this challenge, recent studies have shifted focus toward analyzing NOMA performance under fading and Im-N environments [178]. To address this gap, we investigate the performance of NOMA-integrated HCNs modeled using PPP under both AWGN and Im-N conditions. This study offers a comparative performance evaluation across noise models, emphasizing the robustness and limitations of NOMA in the presence of impulsive interference—an aspect that, to the best of our knowledge, has not been thoroughly explored in existing literature. This chapter presents several noteworthy contributions, detailed as follows.

- To maintain mathematical tractability while accurately representing the spatial characteristics of the HCN, we employed the PPP model to characterize the spatial distribution of both MBSs and PBSs. This approach ensures spatial consistency in the network modeling, allowing for realistic and analytically manageable representations of base station locations.
- The study undertakes a comparative analysis between the HCN employing NOMA and the traditional OMA systems under the influence of Im-N. In this context, we derived closed-form expressions for the OP of both systems, considering two distinct noise environments: AWGN and impulsive noise. These analytical results provide valuable insights into the resilience and reliability of both access schemes under challenging noise conditions.
- We investigate the combined effect of two key impulsive noise parameters: the impulse noise occurrence probability and the impulsive-to-Gaussian noise power ratio. By analyzing their interplay, this study elucidates how these factors jointly influence the overall system performance, highlighting critical aspects that affect network reliability in impulsive noise environments.

- Extensive simulation results substantiate the analytical findings and demonstrate that the HCN-NOMA system consistently achieves superior spectral efficiency and energy efficiency compared to the OMA system. This performance advantage is observed under both AWGN and impulsive noise scenarios, underscoring the robustness and effectiveness of the proposed NOMA-based heterogeneous network design.

5.1 System Model

A two-tier HCN, integrated NOMA with MBS and NOMA-enabled PBS enhances system flexibility, accommodating users with distinct channel conditions and diverse QoS requirements in the shared downlink spectrum is shown in Fig. 5.1. PPP is utilized in HCN for the mathematical modeling of MBS and PBS distributions, offering a spatially realistic framework for wireless network performance evaluation and system design. Nodes and BSs are independently distributed using PPP (Ω_t, λ_t) for tiers $(t \in \{m, p\})$, and users are distributed via PPP Ω_u with density λ_u [146]. Without loss of generality, each UE, or BS, is equipped with a single antenna and operates in half-duplex mode.

The connectivity between a MBS and a macro-based user (MU) adheres to the nearest neighbor connection policy. A bounded path loss model denoted by $P(r) = \frac{1}{1+r_t^{\nu_t}}$ [146] is considered. The path loss exponent for the tth tier is denoted by ν_t, with r_t representing the distance between the serving BS and the typical user. Consequently, $|h_t|^2 = |\tilde{h}_t|^2 P(r)$ defines the total channel gain experienced by the selected user in the tth tier, where $\tilde{h}_t \sim \mathcal{CN}(0, 1)$. The system-wide transmission bandwidth is set 1 Hz, and R denotes the desired data rate for the specific user. The $\tilde{h}_t$ represents the actual complex Gaussian distributed channel, where $|h|^2$ and $|\ddot{h}|^2$ denotes ordered and unordered channel gains, respectively.

5.1.1 *PBS User's SINR (without NOMA)*

Mathematical expression of the signal transmitted from the PBS is denoted as $X_{p,tx} = x_p\sqrt{P_p}$, where x_p represents the intended signal to the specific user linked to PBS (PU), and P_p denotes the transmit power of the PBS tier. While the signal received at the PU is represented as

$$X_{p,rx} = \sqrt{P_p}\ddot{h}_{p_o}x_p + \underbrace{\sum_{t\in(m,p)} \sqrt{P_t}x_t \sum_{r\in\Omega_t\setminus\{p_o\}} \ddot{h}_r}_{\text{Intra-tier and Inter-tier interference by BSs}} + n_p, \tag{5.1}$$

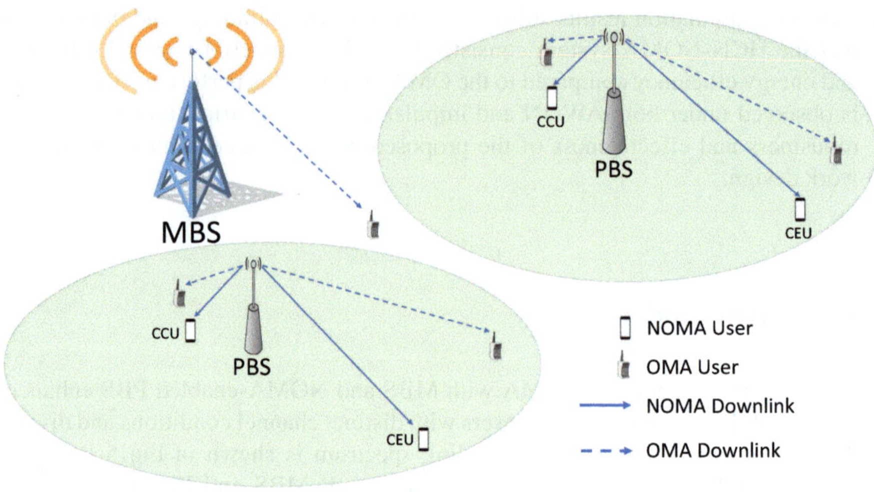

Fig. 5.1 System model of HCN NOMA

where the total noise n_p, modeled by the Bernoulli-Gaussian process, frequently employed in the performance analysis and design of communication systems for its simplicity and effectiveness in characterizing random occurrence of high-power impulses [175], is expressed as

$$n_p = n_w + \delta \times n_I \tag{5.2}$$

where n_w represents the background noise, and n_I represents the impulsive noise such that $n_w \sim \mathcal{CN}(0, \sigma_w^2)$, and $n_I \sim \mathcal{CN}(0, \sigma_I^2)$. The parameter "$\delta$" follows a Bernoulli process, characterized by an independent and identically distributed (i.i.d.) 0s and 1s sequence. It takes the value "1" with probability "p" and "0" with probability "$1 - p$". The PDF for the overall noise n_p is detailed in [175, 178], is expressed as,

$$f(n_p) = \frac{(1-p)}{\pi \sigma_w^2} e^{-\frac{|n_p|^2}{\sigma_w^2}} + \frac{p}{\pi(\sigma_w^2 + \sigma_I^2)} e^{-\frac{|n_p|^2}{\sigma_w^2 + \sigma_I^2}} \tag{5.3}$$

The average noise power ζ is given by [179]

$$\begin{aligned} \zeta &= \mathbb{E}[n_p^2] = \mathbb{E}[n_w^2] + \mathbb{E}[\delta^2] \times \mathbb{E}[n_I^2] \\ &= \sigma_w^2 + p \times \sigma_I^2 = \sigma_w^2 \times \zeta_p \end{aligned} \tag{5.4}$$

where ζ_p is given as $1 + p\lambda$ such that $\lambda = \frac{\sigma_I^2}{\sigma_w^2}$ quantifies the impulsive-to-Gaussian noise power ratio. The SINR at a selected PU can be expressed as [146]

Table 5.1 Comparative analysis with related studies

	[114]	[109]	[146]	[178]	[180]	[181]	This work
	HCN+NOMA			NOMA+Im-N			
HCN multi-tier analysis	✓	✓	✓	✗	✗	✗	✓
NOMA	✓	✓	✓	✓	✓	✓	✓
PPP distributions	✓	✓	✓	✗	✗	✗	✓
OP/BER	✓	✓	✓	✓	✓	✓	✓
System Throughput	✗	✗	✓	✗	✗	✗	✓
Energy Efficiency	✗	✗	✓	✗	✗	✗	✓
Im-N environment	✗	✗	✗	✓	✓	✓	✓
Compare: OMA/NOMA	✓	✓	✓	✗	✓	✗	✓
Compare: AWGN/Im-N	✗	✗	✗	✓	✗	✓	✓

$$\gamma_p = \frac{P_p|\ddot{h}^{p_o}|^2}{\sum_t P_t^\iota \mathbb{I}_t + \sigma_w^2 + p\sigma_I^2} = \frac{\rho_p P_p|\ddot{h}^{p_o}|^2}{\sum_t \rho_t^\iota \mathbb{I}_t + \zeta_p} \tag{5.5}$$

where ρ_p is the transmit SNR at the PBS tier and is determined as $\rho_p = \mathbb{E}\left[x_p^2\right]/\sigma_w^2$. The transmit SNR from the tth tier, which interferes with the selected PU is given as $\rho_t^\iota = P_t/\sigma_w^2$. Without loss of generality, if the selected PU is situated at the point of origin and the designated PBS is positioned at p_o, according to Slivnyak's theorem [133], $\rho_p^\iota \mathbb{I}_p$ denotes the intra-tier interference caused by the PBS tier on the PU. In this case, $\mathbb{I}_p = \sum_{j\in\Omega_p\setminus\{p_o\}} |\ddot{h}_j|^2$, where $|\ddot{h}_j|^2$ is the overall channel gain from the jth PBS to the selected PU. Likewise, $\rho_m^\iota \mathbb{I}_m$ denotes inter-tier interference originating from the MBS tier towards the selected PU. In such case, $\mathbb{I}_m = \sum_{i\in\Omega_m} |\ddot{h}_i|^2$, where $|\ddot{h}_i|^2$ represents the total channel gain for the selected PU from the ith transmitter of the MBS tier (Table 5.1).

5.1.2 *PBS User's SINR (with NOMA)*

The power domain NOMA is applied by the PBS to support M_p users, where the power allocation factors $a_1 \geq \cdots \geq a_{M_p}$ are determined according to the sorted channel gains $|h_1^{p_o}|^2 \leq \cdots \leq |h_{M_p}^{p_o}|^2$ of the NOMA users. In NOMA, a_n signifies the power coefficients for the nth user. In this context, a 'near user' or 'CCU' is defined by having a strong channel gain, whereas a 'far user' or 'CEU' has a weaker channel gain. Assuming uniform signal power for all NOMA users, the ith NOMA user's intended signal is expressed as x_i, while the transmitted signal by the PBS is represented by $X_{p,tx} = \sum_{i=1}^{M_p} x_i\sqrt{a_i P_p}$. Without loss of generality, we assume that $\mathbb{E}\left[|x_i|^2 = 1\right]$. The signal received by the kth user is expressed as

$$X_{k,rx}^{p} = h_k^{p_o} s_k \sqrt{P_p} + \underbrace{\sum_{t\in(m,p)} x_t \sqrt{P_t} \sum_{r\in\Omega_t\setminus\{p_o\}} h_r}_{\text{Intra-tier and Inter-tier interference by BSs}} + n_k, \tag{5.6}$$

where s_k comprises of $s_k = \sqrt{a_k} x_k + \sum_{u=k+1}^{M_p} \sqrt{a_u} x_u$. The k-th user performs SIC by decoding the message of the jth user; thus, the SINR at the kth user, where $j \leq k$, is denoted as

$$\gamma_{j\to k}^{p} = \frac{\rho_p P_p a_j |h_k^{p_o}|^2}{\rho_p P_p |h_k^{p_o}|^2 \left(\sum_{n=j+1}^{M_p} a_n\right) + \sum_t \rho_t^t \mathbb{I}_t + \zeta_p} \tag{5.7}$$

and a_n is the power allocation factor for users indexed by $n \in \{k, j\}$. The user at the maximum distance from the BS remains uninfluenced by i-SIC, since no SIC is performed, treating signals from other users as noise. Expressing the SINR for the kth user decoding its own message as

$$\gamma_k^{p} = \frac{\rho_p P_p a_k |h_k^{p_o}|^2}{\rho_p P_p |h_k^{p_o}|^2 \left(\sum_{n=k+1}^{M_p} a_n\right) + \sum_t \rho_t^t \mathbb{I}_t + \zeta_p}. \tag{5.8}$$

5.2 Performance Analysis

In this section, the rate OP is assessed across various tiers for the designated systems outlined in Sect. 5.1. OP quantifies the likelihood of performance degradation in a wireless link, such as signal loss, lower data rates, or increased errors.

5.2.1 *Outage Analysis for PBS Tier (OMA)*

The rate OP at a typical PU is determined by considering a uniform distance from the PBS, is given by

$$\begin{aligned} \mathbb{P}_O^p &= \Pr\left[W_p \times \log_2(1+\gamma_p) < R\right], \\ &= \Pr\left[|\ddot{h}_{p_o}|^2 < \frac{\phi_p}{\rho_p P_p}\left(\sum_t \rho_t^t \mathbb{I}_t + \zeta_p\right)\right], \end{aligned} \tag{5.9}$$

where W_p represents the bandwidth assigned to OMA-based PU, and γ_p denotes the SINR at PU, as defined in (5.5), and R indicates the user's desired data rate. Consequently, the SINR threshold is defined as $\phi_p = 2^{2R} - 1$. Thus, the OP can be expressed as,

$$\mathbb{P}_O^p \approx \sum_{q=0}^{Q} b_q^p e^{-\frac{c_q^p \phi_p \zeta_p}{\rho_p P_p}} \prod_t \mathcal{L}_{\mathbb{I}_t}(\rho_t^l s_p), \tag{5.10}$$

where Q strikes a balance between accuracy and complexity, $b_q^p = -w_Q \sqrt{1-\theta_q^2}$ $\left(\frac{1}{2}\left(\theta_q+1\right)\right), w_Q = \frac{\pi}{Q}, \theta_q = \cos\left(\frac{2q-1}{2Q}\pi\right), b_0 = -\sum_{q=1}^{Q} b_q^p, c_q^p = 1 + \left(\frac{\mathcal{Y}_p}{2}\theta_q + \frac{\mathcal{Y}_p}{2}\right)_p^{\nu},$ $c_0 = 0$, and $s_p = \frac{c_q^p \phi_p}{\rho_p P_p}$, such that $\mathcal{Y}_p$ denotes coverage range of PBS tier.

LT of interference from the tth tier is $\mathcal{L}_{\mathbb{I}_t}(s)$ expressed [146] as

$$\mathcal{L}_{\mathbb{I}_t}(s) = e^{\pi \lambda_t (s^{\mu_t} \Gamma(1-\mu_t, s) - s^{\mu_t} \Gamma(1-\mu_t))}, \tag{5.11}$$

where $\mu_t = 2/\nu_t$, $\Gamma(z) = \int_0^\infty x^{z-1} e^{-x} dx$ and $\Gamma(z, x) = \int_x^\infty t^{z-1} e^{-t} dt$.

High SNR Approximation: High SNR asymptotic analysis clarifies system behavior without noise or interference, highlighting dominant mathematical terms' significance. When $\rho_p \to \infty$, a high SNR approximation of (5.10) with $e^{-x} \approx 1 - x$ for every value of x very close to zero is given by

$$\mathbb{P}_O^{p,\infty} \approx -\sum_{q=0}^{Q} b_q^p \frac{c_q^p \phi_p \zeta_p}{\rho_p P_p} \prod_t \mathcal{L}_{\mathbb{I}_t}(\rho_t^l s_p). \tag{5.12}$$

Proof Please refer to Appendix C.1. □

5.2.2 *Outage Analysis for PBS Tier (NOMA)*

We examine a scenario where the user's QoS requirements specify a predefined target data rate R, leading to the formulation of the OP as follows:

$$\mathbb{P}_k^{O,p} = \Pr\left[\gamma_k^p < \phi_k, \gamma_{j\to k}^p < \phi_j\right], \tag{5.13}$$

where $\gamma_{j\to k}^p$ and γ_k^p are given in (5.7), and (5.8) respectively.

Given the uniform distance from the PBS and considering the ordered channel gains, the rate OP at the kth selected PU is given by

$$\mathbb{P}_k^{O,p} \approx \epsilon_k \sum_{d=0}^{M_p-k} \binom{M_p-k}{d} \frac{(-1)^d}{k+d} \sum_{T_k^d} \binom{k+d}{l_0+\ldots+l_N} \prod_{q=0}^{Q} (b_q^p)^{l_q}$$
$$\times e^{-\sum_{q=0}^{Q} l_q \frac{c_q^p \psi_{\max} \zeta_p}{\rho_p P_p}} \prod_t \mathcal{L}_{\mathbb{I}_t}(\rho_t^l s_p), \tag{5.14}$$

where $\epsilon_k = \frac{M_p!}{(k-1)!(M_p-k)!}$, $s_p = \frac{\psi_{\max} \sum_{q=0}^{Q} l_q c_q^p}{\rho_p P_p}$, $T_k^d = \left\{(l_0, ..., l_N) | \sum_{i=0}^{N} l_i = k + d\right\}$, and $\binom{k+d}{l_0+...+l_N} = \frac{(k+d)!}{l_0!...l_N!}$.

High SNR Approximation: When $\rho_p \to \infty$, a high SNR approximation of (5.14) with $e^{-x} \approx 1 - x$ for every value of x very close to zero is given by

$$\mathbb{P}_{k,\infty}^{O,p} \approx \epsilon_k \sum_{d=0}^{M_p-k} \binom{M_p - k}{d} \frac{(-1)^d}{k+d} \sum_{T_k^d} \binom{k+d}{l_0 + ... + l_N} \prod_{q=0}^{Q} (b_q^p)^{l_q} \times \left(1 - \sum_{q=0}^{Q} l_q \frac{c_q^p \psi_{\max} \zeta_p}{\rho_p P_p}\right) \prod_t \mathcal{L}_{\mathbb{I}_t}(\rho_t^l s_p). \tag{5.15}$$

Proof Please refer to Appendix C.2.

□

5.3 Research Findings and Analysis

In Sect. 5.2, analytical expressions for the OP of both MU and PBS tier users for both OMA and NOMA environments subject to impulsive noise have been derived. In this section, inferences are drawn based on the simulation results on the OP, from which system throughput and energy efficiency are derived. The analytical results are validated using simulations, with parameters including MBS transmitting power (P_m) set at 40 W and PBS transmitting power (P_p) at 1 W, with MBS node density (λ_m) at $5 \times 10^{-5}\text{m}^{-2}$, PBS node density ($\lambda_p$) at 10^{-4}m^{-2}, and user node density (λ_u) at $5 \times 10^{-4}\text{m}^{-2}$. Power allocation coefficients for NOMA users (a_k) are 0.8 and 0.2. Coverage ranges are set to 1000 m for MBS ($\mathcal{Y}_m$) and 5 m for PBS ($\mathcal{Y}_p$), with path loss exponents of 3 for MBS (ν_m) and 4 for PBS (ν_p). Target data rates for CEU (R_1) and CCU (R_2) are specified as in the respective figures, similar to the ones used by [146, 172, 178].

Figure 5.2 illustrates the influence of impulsive noise on the outage performance of the HCN-NOMA system specifically within the PBS tier, considering a scenario involving two users. The figure depicts the OP plotted against different values of impulsive noise occurrence probabilities for users served by NOMA-enabled PBS, alongside a benchmark comparison with the conventional AWGN environment. The results reveal a pronounced deterioration in outage performance for both CEUs and CCUs when exposed to impulsive noise, as opposed to the relatively more benign AWGN conditions. Particularly for CCUs, a notable improvement in performance is observed as the probability of impulsive noise occurrence decreases: reducing the impulsive noise probability from 0.1 to 0.01 and eventually to 0 leads to gains of approximately 2.617 and 0.393 dB, respectively, when the OP reaches the level of 10^{-2}. This indicates a meaningful reduction in the likelihood of communication failure as the impulsive noise environment becomes less severe. A similar behavior

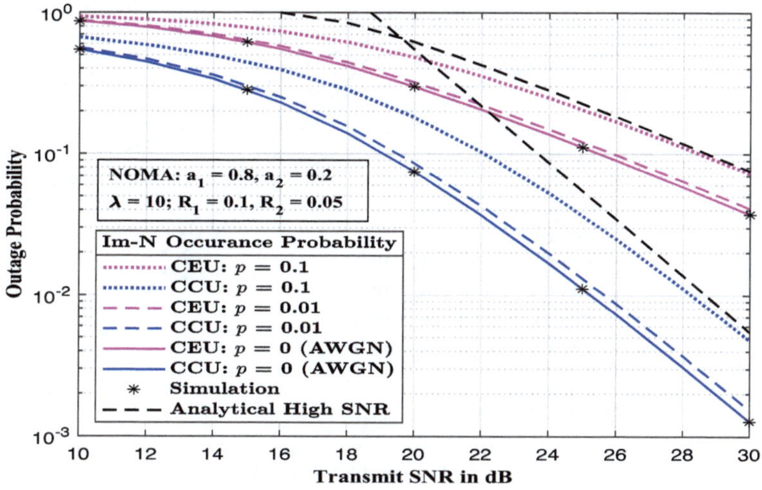

Fig. 5.2 OP across multiple occurrence probability of impulse noise utilizing NOMA

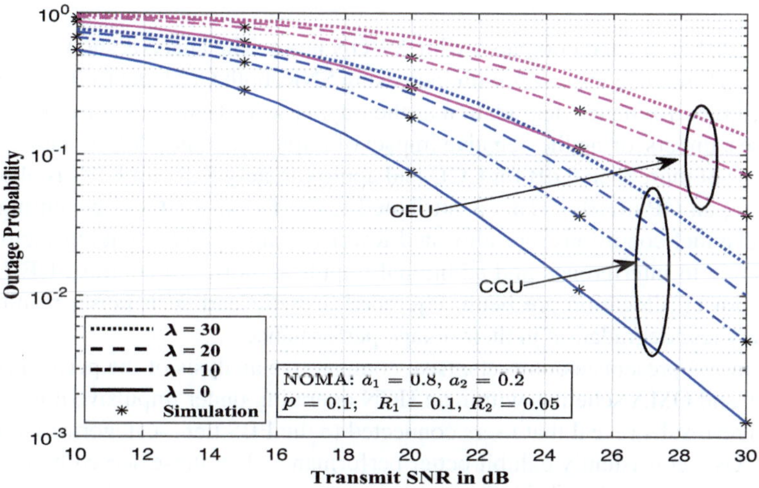

Fig. 5.3 OP across different impulsive to Gaussian noise power ratio utilizing NOMA

is seen for CEUs, where a decline in the impulsive noise occurrence probability correlates with enhanced outage performance, reinforcing the sensitivity of both user categories to the presence of impulsive noise. These observations underscore the critical need to consider and effectively mitigate impulsive noise effects in the design and implementation of NOMA systems within HCNs, to ensure reliable and robust communication performance.

The OP as a function of varying impulsive-to-Gaussian noise power ratios, with a fixed occurrence probability of impulsive noise, is depicted in Figure 5.3 for NOMA-

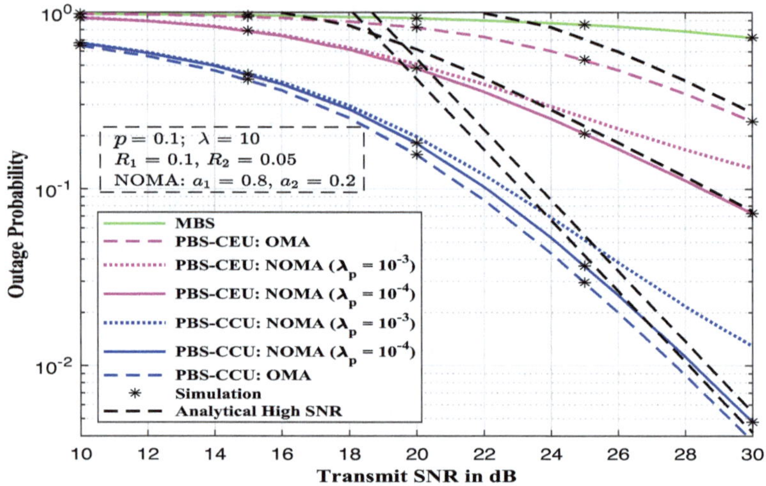

Fig. 5.4 OP for both OMA and NOMA subject to impulsive noise

enabled PBS users and compared with AWGN noise. The analysis shows that while outage performance improves with increasing SNR, both CEU and CCU experience significant performance degradation as the impulsive-to-Gaussian noise power ratio increases, underscoring the negative impact of impulsive noise. Additionally, at $\lambda = 30$, a loss of more than 6 dB for CCU and approximately 5 dB for CEU is observed, achieving an OP of 10^{-1}. This is attributed to the fact that CCUs, benefiting from better channel conditions, are allocated lower transmit power, making them more susceptible to interference and additional impulsive noise. In contrast, CEUs with higher power allocation are better equipped to withstand such noise, resulting in relatively less degradation in their outage performance.

Figure 5.4 presents a comprehensive comparative analysis of the performance of NOMA and OMA schemes within an HCN operating under impulsive noise conditions. The results reveal that users connected to the PBS tier—categorized as CEUs and CCUs—consistently exhibit better performance than those served by the MBS tier. Within the PBS tier, CCUs demonstrate superior performance compared to CEUs across both access methods. NOMA offers a distinct advantage for CEUs, enhancing their outage performance, whereas OMA tends to deliver better results for CCUs, particularly when certain power allocation coefficients are employed. This effect is especially pronounced when considering variations in impulsive noise power ratios and occurrence probabilities, highlighting the critical role of power allocation strategies in optimizing system performance under such noise conditions. Furthermore, the analysis indicates that an increased density of MBSs positively impacts the outage probability of macro users by reducing the average distance between users and MBSs. Conversely, a lower density of PBSs benefits NOMA users by minimizing interference within the proposed HCN-NOMA framework, thus improving their performance. Figures 5.2 and 5.4 also illustrate the close agreement between the

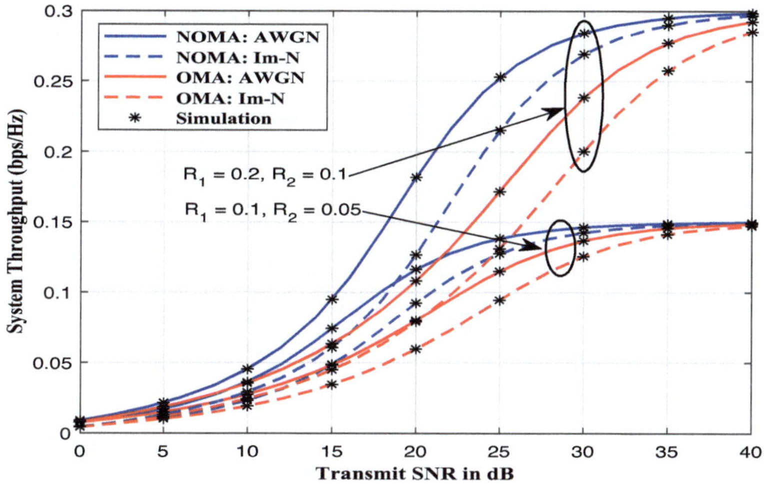

Fig. 5.5 Spectral efficiency in HCN using OMA and NOMA subject to impulsive noise

asymptotic and simulated outage probabilities at high SNR values. This consistency validates the accuracy of the derived closed-form expressions given in equations (5.12) and (5.15), confirming the reliability of the analytical models employed in this study.

System throughput is calculated as $\tau_{sys} = \left(1 - \mathbb{P}_{CEU}^{O,p}\right) R_1 + \left(1 - \mathbb{P}_{CCU}^{O,p}\right) R_2$, where the OP at the PBS tier for the CEU, and CCU, denoted as $\mathbb{P}_{CEU}^{O,p}$ and $\mathbb{P}_{CCU}^{O,p}$ respectively, are determined by operating (5.14) at $k = 1$ and $k = 2$, with R_1 and R_2 denoting the target data rate for the CEU and CCU, as per [146, 172]. Figure 5.5 compares the spectral efficiency of NOMA and OMA systems under Im-N and AWGN scenarios across different target data rates. NOMA shows better overall system performance and higher throughput than OMA in both cases. Increasing signal power improves outage performance for CCUs with minimal negative impact on CEUs. However, both NOMA and OMA experience performance degradation under Im-N conditions compared to AWGN. Despite the overall performance degradation in Im-N, NOMA still achieves higher spectral efficiency than OMA under AWGN conditions, particularly for SNR values above 15 dB, for the given simulation parameters.

Energy efficiency, defined as the ratio of system throughput to total energy consumed $\left(\eta_{sys} = \tau_{sys} \div P_p\right)$, is analyzed for both NOMA and OMA systems in Fig. 5.6 under Im-N and AWGN scenarios across different target data rates, where τ_{sys} is system throughput, and P_p is transmit power at the PBS [146]. In the HCN-NOMA system, energy efficiency initially increases with rising SNR, reaching a peak at specific SNR values before declining as SNR continues to increase. This decline occurs because, in the high SNR regime, the network consumes more power relative to the achieved system throughput. The analysis shows that energy efficiency peaks at $\approx 9.1 \times 10^{-3}$ at -1.0 dB for NOMA and $\approx 10.1 \times 10^{-3}$ at -2.0 dB for OMA

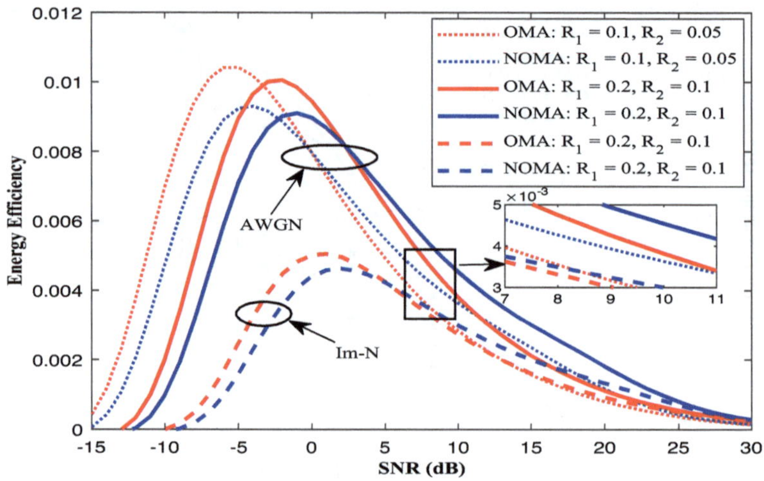

Fig. 5.6 Energy Efficiency in HCN using OMA and NOMA subject to impulsive noise

under the AWGN scenario. Similar trends are observed under the Im-N scenario, though the efficiency values are significantly lower ($\approx 4.63 \times 10^{-3}$ at 2.0 dB for NOMA and $\approx 5.06 \times 10^{-3}$ at 1.0 dB for OMA) compared to AWGN. These peaks are noted when the system targets fixed data rates of 0.2 and 0.1 for CEUs and CCUs, respectively. NOMA consistently outperforms OMA across the positive SNR spectrum, especially at higher SNRs, with a slight shift towards higher peak energy efficiency at lower SNRs for lower target data rates. NOMA outperforms OMA at higher SNRs in the Im-N scenario ($\approx$5.5 dB) compared to AWGN ($\approx$ 2.35 dB) due to the dominant noise in Im-N environment requiring higher SNR for better NOMA performance.

In Fig. 5.7, the analysis shows that NOMA significantly outperforms OMA in terms of spectral efficiency, with gains of 51, 45, 20, and 7% at transmit SNR levels of 15, 20, 25, and 30 dB under AWGN. However, these gain margins fluctuate with varying SNRs, reflecting the non-linear impact of OP on system throughput. Under AWGN, NOMA shows better performance compared to OMA at lower transmit SNRs due to reduced noise impact, while in the Im-N environment as noise becomes the dominant factor, NOMA outperforms at higher transmit SNRs, underscoring the need for operating in higher SNRs under such scenarios.

5.4 Summary

This chapter compares the impact of Im-N in an HCN downlink NOMA system. The findings highlight substantial impact on the outage performance, system throughput, and energy efficiency within the HCN-NOMA system, particularly under different

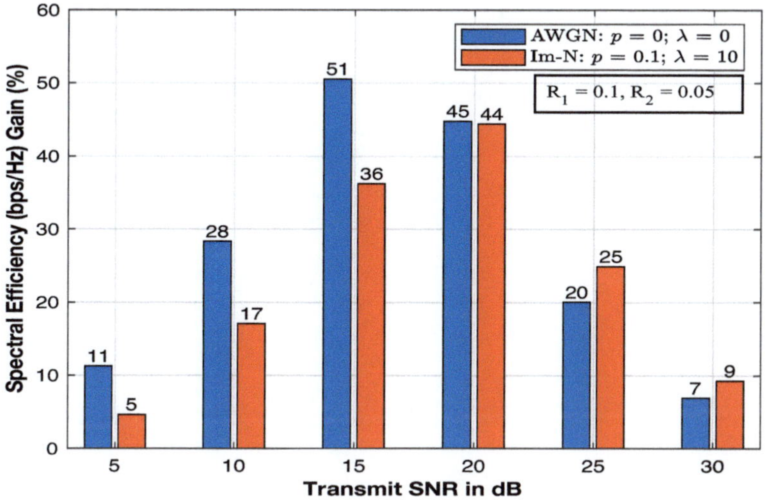

Fig. 5.7 Quantitative analysis: comparing NOMA versus OMA gain (%) subject to impulsive noise

data rate scenarios. Numerical findings show NOMA's superiority over OMA, particularly under AWGN as well as Im-N scenario, across a broad SNR range. The results highlight NOMA's superior performance in AWGN scenarios, especially at lower SNRs; however, its effectiveness decreases in Im-N conditions, underscoring the need for operating in higher SNRs. Increased MBS node density improves OP for macro-based users due to shorter user-to-MBS distances, while decreased PBS node density enhances performance for NOMA users by reducing interference in the proposed HCN-NOMA system. This analysis emphasizes the need for careful decision-making in deploying NOMA techniques in HCN under different noise conditions. CSI precision is critical for NOMA's SIC, and future work should examine the impact of imperfect CSI on PPP-distributed users in NOMA systems, considering both perfect and i-SIC within an HCN, using this study as a benchmark.

Chapter 6
Active RIS-Aided NOMA Network

The investigation of HCN-NOMA under impulsive noise in Chap. 5 sets the stage for exploring advanced techniques to further enhance network performance, leading to the study of reconfigurable intelligent surface (RIS)-assisted NOMA in this chapter. While Chapter 5 addresses the challenges posed by impulsive electromagnetic interference and demonstrates NOMA's efficiency over OMA, it also highlights the need for innovative solutions to improve system robustness. This motivates the introduction of RIS in this chapter, where intelligently controlled reflective surfaces are leveraged to optimize signal quality, mitigate interference, and enhance spectral efficiency. By integrating RIS into the NOMA framework, this chapter extends the discussion on noise resilience and energy efficiency by analyzing key performance metrics such as outage probability, ergodic capacity, and symbol error rate under diverse modulation schemes. The findings from this study reinforce the significance of RIS in future wireless networks, demonstrating its potential to improve link reliability and boost overall network performance.

RIS is a pivotal technology for next-generation wireless communication systems due to its capability to reconfigure the wireless propagation environment. RIS is a planar metasurface consisting of a large number of reflecting elements (REs). The phase and/or amplitude of the incident signal are manipulated by a RIS controller, enabling precise reflective beamforming [182]. A passive RIS consists of numerous passive REs, each capable of reflecting the incident signal with a tunable phase shift. However, it encounters a multiplicative fading effect due to the path loss in the transmitter-RIS-receiver link. This effect arises from the combined path losses in the transmitter-RIS and RIS-receiver links [183]. To address this, an active RIS can be employed, which amplifies the reflected signal by substituting passive REs with active REs [184]. Active REs are equipped with active reflection-type amplifiers, such as asymmetric current mirrors and current-inverting converters [185]. As a result, an active RIS boosts signal strength, improving both coverage and spectral efficiency, provided that the thermal noise introduced by the active REs remains within acceptable limits. Integrating active RIS with NOMA can significantly enhance NOMA

V. Bhatia et al., *Advanced NOMA Techniques for Heterogeneous Cellular Networks*, Wireless Networks, https://doi.org/10.1007/978-3-032-09551-0_6

performance by dynamically reconfiguring the wireless propagation environment. This is especially advantageous for mobile users encountering obstacles that may disrupt their connectivity. Therefore, this chapter examines the performance of an active RIS-assisted NOMA network. In [186–191], a passive RIS-aided NOMA networks are explored. In [186], a straightforward design for an RIS-assisted downlink NOMA network is proposed, where the RIS is positioned closer to the cell-edge user, assuming the absence of a direct link between the BS and the cell-edge user. In [187], the influence of phase shifts on the performance of a RIS-assisted NOMA network is analyzed for both coherent and discrete phase shift designs. The study explores the trade-offs between system performance and complexity associated with these designs. In [188], the performance of RIS-assisted NOMA and OMA networks is examined for both downlink and uplink transmissions over Nakagami-m fading channels. In [189], a RIS-assisted downlink NOMA network is introduced for a multi-user two-cell scenario. The RIS is strategically placed at the boundary of two adjacent cells to facilitate joint transmission in a coordinated multipoint setup. In [190], the performance of a RIS-assisted downlink NOMA network is analyzed, considering the effects of phase shift errors, transceiver hardware impairments, and imperfect successive interference cancellation. In [191], a RIS-assisted cooperative NOMA network is proposed, where the NU acts as a relay node to enable device-to-device communication with the FU. The RIS operates using time division multiple access to facilitate communication between the BS and NU while simultaneously improving device-to-device communication between the NU and FU. In [192], the ASER analysis of a passive RIS-assisted full-duplex two-way relay network over Rayleigh fading channels is conducted for rectangular quadrature amplitude modulation (QAM), cross QAM, and hexagonal QAM schemes. In [193], the performance of active RIS-assisted NOMA network with transmit antenna selection is explored. In [194], the joint optimization of phase shift and amplification matrices for an assisting active RIS is explored in a multi-user downlink NOMA network. In [195], the physical-layer security of an active RIS-assisted NOMA network with discrete RIS phase shifts is examined to protect against eavesdropping. In [196], the performance of an active RIS-assisted NOMA network with hardware impairments and SIC imperfections is analyzed. Closed-form expressions for OP and ergodic rate are derived in the study.

Most research articles have laid a solid foundation for passive RIS-assisted NOMA networks. However, passive RIS is affected by the multiplicative path-loss effect. Therefore, in this chapter, we examine the performance of an active RIS-assisted NOMA network that mitigates this path-loss. The key contributions of this work are as follows:

- The analytical expression for OP is derived in terms of the Meijer-G function for both the NU and FU. These formulations account for factors such as the number of REs, amplification gain, thermal noise introduced by the active REs, fading severity, and other system parameters. Additionally, we derive a closed-form expression for the asymptotic OP at high SNR for both the NU and FU, enabling the determination of the diversity order for the network. We demonstrate that the diversity order is affected by the increase in the number of REs and the fading severity, which are driven by fine-grained reflect beamforming and improved channel conditions for signal transmission, respectively.

- The analytical expression for the ergodic capacity of the NU is derived using the Meijer-G function. Furthermore, the ergodic capacity for the FU is derived using the Gaussian Chebyshev quadrature (GCQ) approximation method. The effects of fading severity, the number of REs, amplification gain, and thermal noise are incorporated into the analysis.
- The analytical expression for ASER is derived for both the NU and FU, considering generalized rectangular QAM, cross QAM, and hexagonal QAM schemes. For the FU, the GCQ approximation method is applied. The effects of constellation size, fading severity, the number of REs, amplification gain, thermal noise, and other system parameters are highlighted. Additionally, a performance comparison of ASER is presented for the various QAM schemes, demonstrating the superior performance of hexagonal QAM.
- Finally, Monte Carlo simulations are conducted to validate the accuracy of the derived closed-form expressions, and valuable insights are gained. The performance of the considered NOMA network is compared with that of an OMA network. Increasing the number of REs, amplification gain, and improving fading statistics results in enhanced performance. However, a noticeable performance degradation occurs as the thermal noise power of the active REs increases.

6.1 System Model

In this section, an active RIS-aided downlink NOMA network consisting of a BS (s), active RIS, NU (n), and FU (f) is shown in Fig. 6.1. Specifically, the NU is a cell-center user with a direct link to the BS, while the FU is a cell edge user affected by shadowing and lacking a direct link to the BS. As a result, the FU requires assistance from the active RIS for communication. The BS, NU, and FU each have a single antenna. The active RIS consists of R REs, with each RE equipped with a dedicated active reflection-type amplifier, such as an asymmetric current mirror, current-inverting converters, or specific integrated circuits [185], along with a phase-shift circuit. The reflection-coefficient matrix of the active RIS is represented by $\boldsymbol{\Theta} = \text{diag}(\alpha_1 e^{j\theta_1}, ..., \alpha_R e^{j\theta_R})$, where $\alpha_r > 1$ is the amplification gain and $\theta_r \in [0, 2\pi)$ is the phase-shift of the rth RE with $\text{r} = 1, 2, \ldots, R$. Furthermore, $h_{\text{sn}} \in \mathbb{C}$, $\boldsymbol{h}_{\text{sr}} \in \mathbb{C}^{1\times R}$, and $\boldsymbol{h}_{\text{rf}} \in \mathbb{C}^{R\times 1}$ denote the channel coefficients between the BS and the NU, the BS and the rth RE of the active RIS, and the rth RE of the active RIS and the FU, respectively. It is considered that all channels shows quasi-static behavior, and perfect channel state information at the BS. The channel coefficients experience Nakagami-m fading with m representing the fading severity parameter. Specifically, fading severity of the BS-NU, BS-rth RE of the active RIS, and rth RE of the active RIS-FU links are represented as m_{sn}, m_{sr}, and m_{rf}, respectively. Furthermore, Ω_{sn}, Ω_{sr}, and

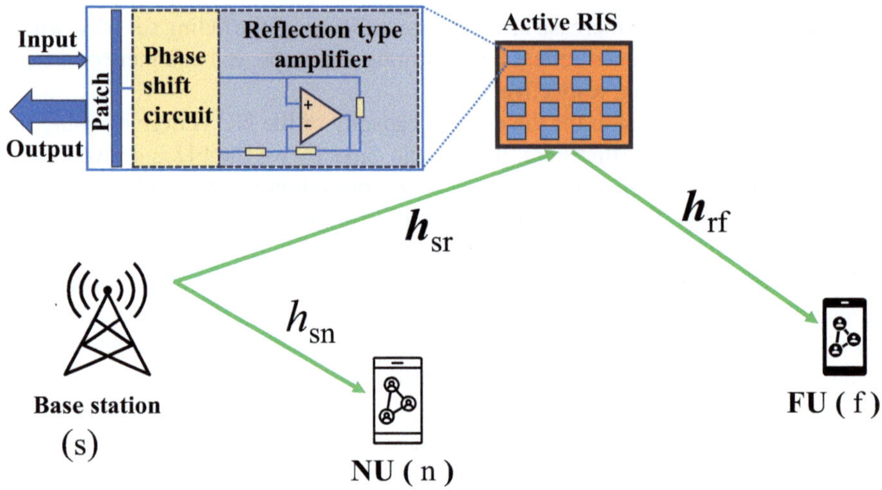

Fig. 6.1 Active RIS-aided NOMA network

Ω_{rf} denote the average power of the BS-NU, BS-rth RE of the active RIS, and rth RE of the active RIS-FU links, respectively.

6.1.1 Signal Model

The BS transmits combined information signals to both the NU and FU using power domain multiplexing and superposition coding. The transmitted signal is given as $x_s = \sqrt{\beta_n} x_n + \sqrt{\beta_f} x_f$, where x_n and x_f represent the transmitted signals to the NU and the FU, respectively. The factors β_n and β_f denote the power allocation coefficients for the NU and FU, respectively, and $\beta_n + \beta_f = 1$. Note that a fixed power allocation shared between the NU and FU is considered, and set $\beta_n < \beta_f$ [186, 187]. The received signals at the NU and the FU are given respectively by

$$y_n = \sqrt{P_s} h_{sn} x_s + w_n, \tag{6.1}$$

$$y_f = \sqrt{P_s} \boldsymbol{h}_{sr} \boldsymbol{\Theta} \boldsymbol{h}_{rf} x_s + \boldsymbol{h}_{rf} \boldsymbol{\Theta} \boldsymbol{w}_r + w_f, \tag{6.2}$$

where P_s represents the transmit power at the BS. The term $w_n \sim \mathcal{CN}(0, \sigma_n^2)$, $\boldsymbol{w}_r \sim \mathcal{CN}(0, \sigma_r^2 \boldsymbol{I}_R)$, and $w_f \sim \mathcal{CN}(0, \sigma_f^2)$ represent the AWGN at the NU, thermal noise due to active REs, and the AWGN at the FU, respectively.

The FU signal is initially decoded at the NU, and the corresponding received instantaneous SINR is expressed as

$$\Lambda_{\mathrm{n}\to\mathrm{f}} = \frac{\zeta_{\mathrm{n}}\beta_{\mathrm{f}}|h_{\mathrm{sn}}|^2}{\zeta_{\mathrm{n}}\beta_{\mathrm{n}}|h_{\mathrm{sn}}|^2+1}, \tag{6.3}$$

where $\zeta_{\mathrm{n}} = \frac{P_s}{\sigma_{\mathrm{n}}^2}$. After the implementation of SIC, the NU's signal is decoded, and the corresponding received instantaneous SNR is expressed as

$$\Lambda_{\mathrm{n}} = \zeta_{\mathrm{n}}\beta_{\mathrm{n}}|h_{\mathrm{sn}}|^2. \tag{6.4}$$

At the FU, its signal is decoded directly while treating the NU's signal as interference, and the corresponding received instantaneous SINR is expressed as

$$\Lambda_{\mathrm{f}} = \frac{\zeta_{\mathrm{f}}\beta_{\mathrm{f}}\alpha^2|\boldsymbol{h}_{\mathrm{sr}}\boldsymbol{\Phi}\boldsymbol{h}_{\mathrm{rf}}|^2}{\zeta_{\mathrm{f}}\beta_{\mathrm{n}}\alpha^2|\boldsymbol{h}_{\mathrm{sr}}\boldsymbol{\Phi}\boldsymbol{h}_{\mathrm{rf}}|^2+\lambda|\boldsymbol{h}_{\mathrm{rf}}\boldsymbol{\Phi}|^2+1}, \tag{6.5}$$

where $\zeta_{\mathrm{f}} = \frac{P_s}{\sigma_{\mathrm{f}}^2}$ and $\lambda = \frac{\alpha^2\sigma_{\mathrm{r}}^2}{\sigma_{\mathrm{f}}^2}$. Here, for simplicity, we consider $\alpha_{\mathrm{r}} = \alpha$ and define $\boldsymbol{\Theta} = \alpha\boldsymbol{\Phi}$, where $\boldsymbol{\Phi} = \mathrm{diag}(e^{j\theta_1}, \ldots, e^{j\theta_R})$.

6.2 Near User Performance Analysis

In this section, closed-form expressions for OP, asymptotic OP, ergodic capacity, and ASER for generalized rectangular QAM, cross QAM, and hexagonal QAM schemes are derived.

6.2.1 Outage Probability

In a transmission system with delay constraints, OP is a key performance metric that represents the probability of link failure. The OP of the NU is expressed as

$$\begin{aligned}
\mathbb{P}_{\mathrm{n}} &= 1-\Pr\left[\Lambda_{\mathrm{n}\to\mathrm{f}}\geq\Lambda_{\mathrm{thf}},\, \Lambda_{\mathrm{n}}\geq\Lambda_{\mathrm{thn}}\right]\\
&= 1-\Pr\left[\frac{\zeta_{\mathrm{n}}\beta_{\mathrm{f}}|h_{\mathrm{sn}}|^2}{\zeta_{\mathrm{n}}\beta_{\mathrm{n}}|h_{\mathrm{sn}}|^2+1}\geq\Lambda_{\mathrm{thf}},\, \zeta_{\mathrm{n}}\beta_{\mathrm{n}}|h_{\mathrm{sn}}|^2\geq\Lambda_{\mathrm{thn}}\right]\\
&= 1-\Pr\left[|h_{\mathrm{sn}}|^2\geq\frac{\Lambda_{\mathrm{thf}}}{\zeta_{\mathrm{n}}\left(\beta_{\mathrm{f}}-\beta_{\mathrm{n}}\Lambda_{\mathrm{thf}}\right)},\, |h_{\mathrm{sn}}|^2\geq\frac{\Lambda_{\mathrm{thn}}}{\zeta_{\mathrm{n}}\beta_{\mathrm{n}}}\right]\\
&= \frac{1}{\Gamma\left(m_{\mathrm{sn}}\right)}\Upsilon\left(m_{\mathrm{sn}}, \frac{m_{\mathrm{sn}}\Xi}{\Omega_{\mathrm{sn}}\zeta_{\mathrm{n}}}\right),
\end{aligned} \tag{6.6}$$

where $\Lambda_{\mathrm{thf}} = 2^{\mathcal{R}_{\mathrm{f}}}-1$, $\Lambda_{\mathrm{thn}} = 2^{\mathcal{R}_{\mathrm{n}}}-1$, and $\Xi = \max\left\{\frac{\Lambda_{\mathrm{thf}}}{\beta_{\mathrm{f}}-\beta_{\mathrm{n}}\Lambda_{\mathrm{thf}}}, \frac{\Lambda_{\mathrm{thn}}}{\beta_{\mathrm{n}}}\right\}$. The terms $\mathcal{R}_{\mathrm{f}}$ and $\mathcal{R}_{\mathrm{n}}$ denote the target rates of the NU and FU, respectively. By employing [197, Eq.

(8.352.6)] and representing $e^{-cx} = \mathrm{G}_{0,1}^{1,0}\left[cx\big|_{0}\right]$ in (6.6), the closed-form expression of OP is obtained as

$$\mathbb{P}_{\mathrm{n}} = 1 - \sum_{p=0}^{m_{\mathrm{sn}}} \frac{1}{p!}\left(\frac{m_{\mathrm{sn}}\Xi}{\Omega_{\mathrm{sn}}\zeta_{\mathrm{n}}}\right)^{p} \mathrm{G}_{0,1}^{1,0}\left[\frac{m_{\mathrm{sn}}\Xi}{\Omega_{\mathrm{sn}}\zeta_{\mathrm{n}}}\bigg|_{0}\right]. \tag{6.7}$$

Additionally, the diversity order of the NU communication link is determined from the asymptotic OP by approximating the OP at high SNR. By using a high SNR approximation $\Upsilon(a, y) \underset{y\to 0}{\approx} (\frac{y^{a}}{a})$ [198] in (6.6), the closed-form expression of the asymptotic OP is obtained as

$$\mathbb{P}_{\mathrm{n}}^{\infty} \approx \mathcal{A}_{\mathrm{n}}(\zeta_{\mathrm{n}})^{-m_{\mathrm{sn}}}, \tag{6.8}$$

where

$$\mathcal{A}_{\mathrm{n}} = \frac{1}{m_{\mathrm{sn}}\Gamma(m_{\mathrm{sn}})}\left(\frac{m_{\mathrm{sn}}\Xi}{\Omega_{\mathrm{sn}}}\right)^{m_{\mathrm{sn}}}. \tag{6.9}$$

From (6.8), the diversity order of the NU is obtained as m_{sn}, i.e., the fading severity parameter of the BS-NU link.

6.2.2 *Ergodic Capacity*

Ergodic capacity, measured in bps/Hz, represents the theoretical upper limit for reliable communication in a fading channel. It is calculated by averaging the instantaneous capacity over fading channels. The ergodic capacity of the NU is expressed as [188, 199]

$$\mathbb{C}_{\mathrm{n}} = \mathbb{E}\left\{\log_2(1+\Lambda_{\mathrm{n}})\right\} = \int_0^{\infty} \log_2(1+\gamma) f_{\Lambda_{\mathrm{n}}}(\gamma)d\gamma. \tag{6.10}$$

Further, (6.10) can be simplified as [200, 201]

$$\mathbb{C}_{\mathrm{n}} = \frac{1}{\ln(2)}\int_0^{\infty} \frac{1}{1+\gamma}\left\{1 - F_{\Lambda_{\mathrm{n}}}(\gamma)\right\}d\gamma. \tag{6.11}$$

The CDF $F_{\Lambda_{\mathrm{n}}}(\gamma)$ is calculated by using (6.4), and is given as

$$F_{\Lambda_{\mathrm{n}}}(\gamma) = 1 - \sum_{p=0}^{m_{\mathrm{sn}}-1} (\alpha_0)^{p}\gamma^{p}e^{-\alpha_0\gamma}, \tag{6.12}$$

where $\alpha_0 = \frac{m_{\mathrm{sn}}}{\Omega_{\mathrm{sn}}\zeta_{\mathrm{n}}\beta_{\mathrm{n}}}$. By substituting (6.12) in (6.11) and representing $\frac{1}{1+\gamma} = \mathrm{G}_{1,1}^{1,1}\left[\gamma\middle|_{0}^{0}\right]$, the closed-form expression of the ergodic capacity for the NU can be evaluated using [202, Eq. (07.34.21.0088.01)] and is given as

$$\mathbb{C}_{\mathrm{n}} = \sum_{p=0}^{m_{\mathrm{sn}}-1} \frac{1}{p!\ln(2)\alpha_0} \mathrm{G}_{2,1}^{1,2}\left[\frac{1}{\alpha_0}\middle|_{0}^{-p,0}\right]. \tag{6.13}$$

6.2.3 Average Symbol Error Rate

The generalized expression for the ASER based on the CDF for any modulation scheme is given as [203, 204]

$$\mathbb{P}_{\mathrm{e,n}} = -\int_0^\infty \mathcal{P}'_s\left(e|\gamma\right) F_{\Lambda_{\mathrm{n}}}(\gamma) d\gamma, \tag{6.14}$$

where $\mathcal{P}'_s\left(e|\gamma\right)$ represents the first-order derivative of the conditional symbol error probability (SEP) $\mathcal{P}_s\left(e|\gamma\right)$.

Rectangular QAM

The conditional SEP of $M_{\mathrm{a}} \times N_{\mathrm{a}}$ rectangular QAM over AWGN channel is expressed as [205, 206]

$$\mathcal{P}_{\mathrm{s}}^{\mathrm{RQAM}}\left(e|\gamma\right) = 2\left[c_1 Q(c_3\sqrt{\gamma})\left(1 - 2c_2 Q\left(c_4\sqrt{\gamma}\right)\right) + c_2 Q\left(c_4\sqrt{\gamma}\right)\right], \tag{6.15}$$

where M_{a} and N_{a} represent the in-phase and quadrature phase constellation points, respectively. The terms $c_1 = 1 - \frac{1}{M_{\mathrm{a}}}$, $c_2 = 1 - \frac{1}{N_{\mathrm{a}}}$, $c_3 = \sqrt{6/\left(\left(M_{\mathrm{a}}^2 - 1\right) + \left(N_a^2 - 1\right)\epsilon\right)}$, $c_4 - \epsilon c_3$, and the ratio $\epsilon = d_Q/d_I$ denotes the distance between quadrature and in-phase decision. Invoking the identity $Q\left(\gamma\right) = 1/2\left[1 - \mathrm{erf}\left(\gamma/\sqrt{2}\right)\right]$ in (6.15) and by using [207, Eq. (7.1.21)], the first-order derivative of the conditional SEP is given as

$$\begin{aligned}\mathcal{P'}_{\mathrm{s}}^{\mathrm{RQAM}}\left(e|\gamma\right) =& \frac{1}{\sqrt{\gamma}}\left[\frac{c_1\left(c_2 - 1\right)c_3}{\sqrt{2\pi}}e^{-\frac{c_3^2\gamma}{2}} + \frac{\left(c_1 - 1\right)c_2c_4}{\sqrt{2\pi}}e^{-\frac{c_4^2\gamma}{2}}\right] \\ &- \frac{c_1c_3c_3c_4}{\pi}e^{-\frac{\left(c_3^2+c_4^2\right)\gamma}{2}}\left[{}_1F_1\left(1; \frac{3}{2}; \frac{c_3^2\gamma}{2}\right) + {}_1F_1\left(1; \frac{3}{2}; \frac{c_4^2\gamma}{2}\right)\right].\end{aligned} \tag{6.16}$$

By substituting (6.16) and (6.12) in (6.14), and invoking the identity ${}_1F_1\left(a; b; x\right) = \sum_{t=0}^{\infty} \frac{(a)_t}{(b)_t t!} x^t$ [197, Eq. (9.14.1)] and by using [197, Eq. (3.381.4)], the ASER expression for rectangular QAM is obtained as

$$\mathbb{P}_{\mathrm{e,n}}^{\mathrm{RQAM}} = -\frac{c_1(c_2-1)c_3}{\sqrt{2\pi}}\left(\frac{c_3^2}{2}\right)^{-\frac{1}{2}}\Gamma\left(\frac{1}{2}\right) - \frac{(c_1-1)c_2c_4}{\sqrt{2\pi}}\left(\frac{c_4^2}{2}\right)^{-\frac{1}{2}}\Gamma\left(\frac{1}{2}\right) + \sum_{t=0}^{\infty}\frac{(1)_t}{\left(\frac{3}{2}\right)_t t!}$$
$$\times \frac{c_1c_2c_3c_4}{\pi}\left\{\left(\frac{c_3^2}{2}\right)^t + \left(\frac{c_4^2}{2}\right)^t\right\}\left(\frac{c_3^2+c_4^2}{2}\right)^{-(t+1)}\Gamma(t+1) + \sum_{p=0}^{m_{\mathrm{sn}}-1}\frac{(\alpha_0)^p}{p!}$$
$$\times \frac{c_1(c_2-1)c_3}{\sqrt{2\pi}}\left(\frac{c_3^2}{2}+\alpha_0\right)^{-\left(p+\frac{1}{2}\right)}\Gamma\left(p+\frac{1}{2}\right) + \sum_{p=0}^{m_{\mathrm{sn}}-1}\frac{(\alpha_0)^p}{p!}\frac{(c_1-1)c_2c_4}{\sqrt{2\pi}}$$
$$\times \left(\frac{c_4^2}{2}+\alpha_0\right)^{-\left(p+\frac{1}{2}\right)}\Gamma\left(p+\frac{1}{2}\right) - \sum_{p=0}^{m_{\mathrm{sn}}-1}\sum_{t=0}^{\infty}\frac{(1)_t}{\left(\frac{3}{2}\right)_t t!}\frac{c_1c_2c_3c_4}{\pi}\frac{(\alpha_0)^p}{p!}$$
$$\times \left\{\left(\frac{c_3^2}{2}\right)^t + \left(\frac{c_4^2}{2}\right)^t\right\}\left(\frac{c_3^2+c_4^2}{2}+\alpha_0\right)^{-(p+t+1)}\Gamma(p+t+1). \tag{6.17}$$

Cross QAM

The conditional SEP of cross QAM over the AWGN channel is given as [208]

$$\mathcal{P}_{\mathrm{s}}^{\mathrm{XQAM}}(e|\gamma) = k_1 Q\left(\sqrt{\frac{2\gamma}{k_2}}\right) + \frac{8}{M_{\mathrm{a}}N_{\mathrm{a}}}\Bigg[\sum_{u=1}^{\frac{k_3}{2}-1} Q\left(2u\sqrt{\frac{2\gamma}{k_2}}\right) + Q\left(k_3\sqrt{\frac{2\gamma}{g}}\right)$$
$$-2\sum_{u=1}^{\frac{k_3}{2}-1} Q\left(\sqrt{\frac{2\gamma}{k_2}}\right)Q\left(2u\sqrt{\frac{2\gamma}{k_2}}\right) - Q\left(\sqrt{\frac{2\gamma}{k_2}}\right)Q\left(k_3\sqrt{\frac{2\gamma}{k_2}}\right)\Bigg]$$
$$-k_4 Q^2\left(\sqrt{\frac{2\gamma}{k_2}}\right), \tag{6.18}$$

where $k_1 = 4 - \frac{2(M_{\mathrm{a}}+N_{\mathrm{a}})}{M_{\mathrm{a}}N_{\mathrm{a}}}$, $k_2 = \frac{2}{3}\left(\frac{31M_{\mathrm{a}}N_{\mathrm{a}}}{32} - 1\right)$, $k_3 = \frac{M_{\mathrm{a}}-N_{\mathrm{a}}}{2}$, and $k_4 = 4 - \frac{4(M_{\mathrm{a}}+N_{\mathrm{a}})}{M_{\mathrm{a}}N_{\mathrm{a}}} + \frac{8}{M_{\mathrm{a}}N_{\mathrm{a}}}$. On invoking the identity $Q(\gamma) = 1/2\left[1 - \mathrm{erf}\left(\gamma/\sqrt{2}\right)\right]$ in (6.18) and using [207, Eq. (7.1.21)], the first-order derivative of conditional SEP is obtained as

$$\mathcal{P}_{\mathrm{s}}^{\prime\mathrm{XQAM}}(e|\gamma) = \frac{1}{2\sqrt{\pi k_2}}\left(-k_1 + \frac{12}{M_{\mathrm{a}}N_{\mathrm{a}}} + k_4\right)\gamma^{-\frac{1}{2}}e^{-\frac{\gamma}{k_2}} - \frac{k_5}{\sqrt{\pi k_2}}\gamma^{-\frac{1}{2}}e^{-\frac{k_3^2\gamma}{k_2}}$$
$$-\sum_{u=1}^{\frac{k_3}{2}-1}\frac{16u}{\pi k_2 M_{\mathrm{a}}N_{\mathrm{a}}}e^{-k_6\gamma}\left\{{}_1F_1\left(1;\frac{3}{2};\frac{\gamma}{k_2}\right) + {}_1F_1\left(1;\frac{3}{2};\frac{4u^2\gamma}{k_2}\right)\right\}$$
$$-\frac{2k_5}{\pi k_2}e^{-\frac{\left(1+k_3^2\right)\gamma}{k_2}}\left\{{}_1F_1\left(1;\frac{3}{2};\frac{\gamma}{k_2}\right) + {}_1F_1\left(1;\frac{3}{2};\frac{k_3\gamma}{k_2}\right)\right\}$$
$$-\frac{k_4}{\pi k_2}e^{-\frac{2\gamma}{k_2}}{}_1F_1\left(1;\frac{3}{2};\frac{\gamma}{k_2}\right), \tag{6.19}$$

where $k_5 = \frac{M_a - N_a}{M_a N_a}$ and $k_6 = \frac{4u^2+1}{k_2}$. By substituting (6.19) and (6.12) in (6.14), and invoking the identity ${}_1F_1(a; b; x) = \sum_{t=0}^{\infty} \frac{(a)_t}{(b)_t t!} x^t$ [197, Eq. (9.14.1)] and solving by using [197, Eq. (3.381.4)], the closed-form expression of ASER for cross QAM is obtained as

$$
\begin{aligned}
\mathbb{P}_{\mathrm{e,n}}^{\mathrm{XQAM}} =& -\frac{1}{2\sqrt{\pi k_2}}\left(-k_1 + \frac{12}{M_a N_a} + k_4\right)\left(\frac{1}{k_2}\right)^{-\frac{1}{2}} \Gamma\left(\frac{1}{2}\right) + \frac{k_5}{\sqrt{\pi k_2}}\left(\frac{k_3^2}{k_2}\right)^{-\frac{1}{2}} \Gamma\left(\frac{1}{2}\right) \\
&+ \sum_{u=1}^{\frac{k_3}{2}-1} \sum_{t=0}^{\infty} \frac{(1)_t}{\left(\frac{3}{2}\right)_t t!} \frac{16u}{\pi k_2 M_a N_a}\left\{\left(\frac{1}{k_2}\right)^t + \left(\frac{4u^2}{k_2}\right)^t\right\}(k_6)^{-(t+1)} \Gamma(t+1) \\
&+ \sum_{t=0}^{\infty} \frac{(1)_t}{\left(\frac{3}{2}\right)_t t!} \frac{2k_5}{\pi k_2}\left\{\left(\frac{1}{k_2}\right)^t + \left(\frac{k_3}{k_2}\right)^t\right\}\left(\frac{1+k_3^2}{k_2}\right)^{-(t+1)} \Gamma(t+1) \\
&+ \sum_{t=0}^{\infty} \frac{(1)_t}{\left(\frac{3}{2}\right)_t t!} \frac{k_4}{\pi k_2}\left(\frac{1}{k_2}\right)^t\left(\frac{2}{k_2}\right)^{-(t+1)} \Gamma(t+1) + \sum_{p=0}^{m_{\mathrm{sn}}-1} \frac{(\alpha_0)^p}{p!} \frac{1}{2\sqrt{\pi k_2}} \\
&\times\left(-k_1 + \frac{12}{M_a N_a} + k_4\right)\left(\frac{1}{k_2} + \alpha_0\right)^{-\left(p+\frac{1}{2}\right)} \Gamma\left(p + \frac{1}{2}\right) - \sum_{p=0}^{m_{\mathrm{sn}}-1} \frac{(\alpha_0)^p}{p!} \frac{k_5}{\sqrt{\pi k_2}} \\
&\times\left(\frac{k_3^2}{k_2} + \alpha_0\right)^{-\left(p+\frac{1}{2}\right)} \Gamma\left(p + \frac{1}{2}\right) - \sum_{p=0}^{m_{\mathrm{sn}}-1} \sum_{u=1}^{\frac{k_3}{2}-1} \sum_{t=0}^{\infty} \frac{(1)_t}{\left(\frac{3}{2}\right)_t t!} \frac{(\alpha_0)^p}{p!} \frac{16u}{\pi k_2 M_a N_a} \\
&\times\left\{\left(\frac{1}{k_2}\right)^t + \left(\frac{4u^2}{k_2}\right)^t\right\}(k_6 + \alpha_0)^{-(p+t+1)} \Gamma(p+t+1) - \sum_{p=0}^{m_{\mathrm{sn}}-1} \sum_{t=0}^{\infty} \frac{(1)_t}{\left(\frac{3}{2}\right)_t t!} \\
&\times \frac{(\alpha_0)^p}{p!} \frac{2k_5}{\pi k_2}\left\{\left(\frac{1}{k_2}\right)^t + \left(\frac{k_3}{k_2}\right)^t\right\}\left(\frac{1+k_3^2}{k_2} + \alpha_0\right)^{-(p+t+1)} \Gamma(p+t+1) \\
&- \sum_{p=0}^{m_{\mathrm{sn}}-1} \sum_{t=0}^{\infty} \frac{(1)_t}{\left(\frac{3}{2}\right)_t t!} \frac{(\alpha_0)^p}{p!} \frac{k_4}{\pi k_2}\left(\frac{1}{k_2}\right)^t\left(\frac{2}{k_2} + \alpha_0\right)^{-(p+t+1)} \Gamma(p+t+1). \quad (6.20)
\end{aligned}
$$

Hexagonal QAM

The conditional SEP of hexagonal QAM in the presence of the AWGN channel is given as [204, 209, 210]

$$
\mathcal{P}_{\mathrm{s}}^{\mathrm{HQAM}}(e|\gamma) = g_1 Q\left(\sqrt{g_3\gamma}\right) + \frac{2}{3} g_2 Q^2\left(\sqrt{\frac{2g_3\gamma}{3}}\right) - 2g_2 Q\left(\sqrt{g_3\gamma}\right) Q\left(\sqrt{\frac{g_3\gamma}{3}}\right), \quad (6.21)
$$

where g_1, g_2, and g_3 are constants, and the values for different constellations in the case of irregular HQAM (optimum) are provided in [203, TAble I]. The first order derivative of (6.21) is expressed as

$$
\begin{aligned}
\mathcal{P}_{\mathrm{s}}^{\prime \mathrm{HQAM}}(e|\gamma) =& \frac{1}{\sqrt{\gamma}}\left[\frac{1}{2}\sqrt{\frac{g_3}{2\pi}}(g_2-g_1)e^{-\frac{g_3\gamma}{2}} - \frac{g_2}{3}\sqrt{\frac{g_3}{3\pi}}e^{-\frac{g_3\gamma}{3}} + \frac{g_2}{2}\sqrt{\frac{g_3}{6\pi}}e^{-\frac{g_3\gamma}{6}}\right] \\
&+ \frac{2g_2g_3}{9\pi}e^{-\frac{2g_3\gamma}{3}}\,{}_1F_1\left(1;\frac{3}{2};\frac{g_3\gamma}{3}\right) \\
&- \frac{g_2g_3}{2\sqrt{3}\pi}e^{-\frac{2g_3\gamma}{3}}\left[{}_1F_1\left(1;\frac{3}{2};\frac{g_3\gamma}{2}\right) + {}_1F_1\left(1;\frac{3}{2};\frac{g_3\gamma}{6}\right)\right].
\end{aligned}
\tag{6.22}
$$

By substituting (6.22) and (6.12) in (6.14), and invoking the identity ${}_1F_1\,(a;b;x) = \sum_{t=0}^{\infty}\frac{(a)_t}{(b)_t t!}x^t$ [197, Eq. (9.14.1)] and by using [197, Eq. (3.381.4)], the ASER expression for hexagonal QAM is obtained as

$$
\begin{aligned}
\mathbb{P}_{\mathrm{e,n}}^{\mathrm{HQAM}} =& -\frac{1}{2}\sqrt{\frac{g_3}{2\pi}}(g_2-g_1)\left(\frac{g_3}{2}\right)^{-\frac{1}{2}}\Gamma\left(\frac{1}{2}\right) + \frac{g_2}{3}\sqrt{\frac{g_3}{3\pi}}\left(\frac{g_3}{3}\right)^{-\frac{1}{2}}\Gamma\left(\frac{1}{2}\right) - \frac{g_2}{2}\sqrt{\frac{g_3}{6\pi}} \\
&\times\left(\frac{g_3}{6}\right)^{-\frac{1}{2}}\Gamma\left(\frac{1}{2}\right) - \sum_{t=0}^{\infty}\frac{(1)_t}{\left(\frac{3}{2}\right)_t t!}\frac{2g_2g_3}{9\pi}\left(\frac{g_3}{3}\right)^t\left(\frac{2g_3}{3}\right)^{-(t+1)}\Gamma(t+1) \\
&+\sum_{t=0}^{\infty}\frac{(1)_t}{\left(\frac{3}{2}\right)_t t!}\frac{g_2g_3}{2\sqrt{3}\pi}\left\{\left(\frac{g_3}{2}\right)^t + \left(\frac{g_3}{6}\right)^t\right\}\left(\frac{2g_3}{3}\right)^{-(t+1)}\Gamma(t+1) + \sum_{p=0}^{m_{\mathrm{sn}}-1}\frac{(\alpha_0)^p}{p!} \\
&\times\frac{1}{2}\sqrt{\frac{g_3}{2\pi}}(g_2-g_1)\left(\frac{g_3}{2}+\alpha_0\right)^{-\left(p+\frac{1}{2}\right)}\Gamma\left(p+\frac{1}{2}\right) - \sum_{p=0}^{m_{\mathrm{sn}}-1}\frac{(\alpha_0)^p}{p!}\frac{g_2}{3}\sqrt{\frac{g_3}{3\pi}} \\
&\times\left(\frac{g_3}{3}+\alpha_0\right)^{-\left(p+\frac{1}{2}\right)}\Gamma\left(p+\frac{1}{2}\right) + \sum_{p=0}^{m_{\mathrm{sn}}-1}\frac{(\alpha_0)^p}{p!}\frac{g_2}{2}\sqrt{\frac{g_3}{6\pi}}\left(\frac{g_3}{6}+\alpha_0\right)^{-\left(p+\frac{1}{2}\right)} \\
&\times\Gamma\left(p+\frac{1}{2}\right) + \sum_{p=0}^{m_{\mathrm{sn}}-1}\sum_{t=0}^{\infty}\frac{(1)_t}{\left(\frac{3}{2}\right)_t t!}\frac{(\alpha_0)^p}{p!}\frac{2g_2g_3}{9\pi}\left(\frac{g_3}{3}\right)^t\left(\frac{2g_3}{3}+\alpha_0\right)^{-(p+t+1)} \\
&\times\Gamma(p+t+1) - \sum_{p=0}^{m_{\mathrm{sn}}-1}\sum_{t=0}^{\infty}\frac{(1)_t}{\left(\frac{3}{2}\right)_t t!}\frac{(\alpha_0)^p}{p!}\frac{g_2g_3}{2\sqrt{3}\pi}\left\{\left(\frac{g_3}{2}\right)^t + \left(\frac{g_3}{6}\right)^t\right\} \\
&\times\left(\frac{2g_3}{3}+\alpha_0\right)^{-(p+t+1)}\Gamma(p+t+1).
\end{aligned}
\tag{6.23}
$$

6.3 Far User Performance Analysis

In this section, we evaluate the performance of the FU by deriving closed-form expressions for OP, asymptotic OP, ergodic capacity, and ASER.

6.3.1 Outage Probability

An outage event at the FU occurs if the received instantaneous SINR, Λ_f (as defined in (6.5)), falls below a predefined threshold Λ_{thf}. For mathematical tractability, Λ_f can be approximated by its lower bound (LB) and upper bound (UB), which are given as

$$\Lambda_f^{LB} \approx \frac{1}{3}\min\left(\frac{\beta_f}{\beta_n}, \frac{\zeta_f\beta_f\alpha^2|\boldsymbol{h}_{sr}\boldsymbol{\Phi}\boldsymbol{h}_{rf}|^2}{\lambda|\boldsymbol{h}_{rf}\boldsymbol{\Phi}|^2}, \zeta_f\beta_f\alpha^2|\boldsymbol{h}_{sr}\boldsymbol{\Phi}\boldsymbol{h}_{rf}|^2\right), \tag{6.24}$$

$$\Lambda_f^{UB} \approx \min\left(\frac{\beta_f}{\beta_n}, \frac{\zeta_f\beta_f\alpha^2|\boldsymbol{h}_{sr}\boldsymbol{\Phi}\boldsymbol{h}_{rf}|^2}{\lambda|\boldsymbol{h}_{rf}\boldsymbol{\Phi}|^2}, \zeta_f\beta_f\alpha^2|\boldsymbol{h}_{sr}\boldsymbol{\Phi}\boldsymbol{h}_{rf}|^2\right). \tag{6.25}$$

The LB of the OP can be evaluated by comparing the UB of the received instantaneous SINR Λ_{act}^{UB} with Λ_{thf}, which can be given as

$$\begin{aligned}
\mathbb{P}_f^{LB} &= \Pr\left[\Lambda_f^{UB} < \Lambda_{thf}\right] \\
&= \Pr\left[\min\left(\frac{\beta_f}{\beta_n}, \frac{\zeta_f\beta_f\alpha^2|\boldsymbol{h}_{sr}\boldsymbol{\Phi}\boldsymbol{h}_{rf}|^2}{\lambda|\boldsymbol{h}_{rf}\boldsymbol{\Phi}|^2}, \zeta_f\beta_f\alpha^2|\boldsymbol{h}_{sr}\boldsymbol{\Phi}\boldsymbol{h}_{rf}|^2\right) < \Lambda_{thf}\right] \\
&= 1 - \mathcal{I}_f\left\{1 - \underbrace{\Pr\left[\frac{\zeta_f\beta_f\alpha^2|\boldsymbol{h}_{sr}\boldsymbol{\Phi}\boldsymbol{h}_{rf}|^2}{\lambda|\boldsymbol{h}_{rf}\boldsymbol{\Phi}|^2} < \Lambda_{thf}\right]}_{\mathcal{I}_1}\right\}\left\{1 - \underbrace{\Pr\left[\zeta_f\beta_f\alpha^2|\boldsymbol{h}_{sr}\boldsymbol{\Phi}\boldsymbol{h}_{rf}|^2 < \Lambda_{thf}\right]}_{\mathcal{I}_2}\right\},
\end{aligned} \tag{6.26}$$

where

$$\mathcal{I}_f = \begin{cases} 1, & \text{if } \beta_f > \beta_n \max(\Lambda_{thf}, 1) \\ 0, & \text{otherwise.} \end{cases} \tag{6.27}$$

Let $Y \triangleq \left(\sum_{r=1}^{R} |h_{sr}||h_{rf}|\right)^2$ for an optimal phase shift (i.e., $\mathrm{ang}(\theta_r) = -\mathrm{ang}(h_{sr}) - \mathrm{ang}(h_{rf})$), where $\mathrm{ang}(\cdot)$ shows the angle. The mean and variance of $\sum_{n=1}^{N} |h_{sr}||h_{rf}|$ are given respectively as

$$\delta = R\frac{\Gamma\left(m_{sr}+\frac{1}{2}\right)\Gamma\left(m_{rf}+\frac{1}{2}\right)}{\Gamma\left(m_{sr}\right)\Gamma\left(m_{rf}\right)}\left(\frac{\Omega_{sr}\Omega_{rf}}{m_{sr}m_{rf}}\right)^{\frac{1}{2}}, \tag{6.28}$$

$$\sigma^2 = R\Omega_{sr}\Omega_{rf}\left\{1 - \frac{1}{m_{sr}m_{rf}}\left(\frac{\Gamma\left(m_{sr}+\frac{1}{2}\right)\Gamma\left(m_{rf}+\frac{1}{2}\right)}{\Gamma\left(m_{sr}\right)\Gamma\left(m_{rf}\right)}\right)^2\right\}. \tag{6.29}$$

By employing the moment-matching method, the PDF and CDF of Y are given as

$$f_Y(y) = \frac{1}{2\Gamma(\mu)(\Delta)^{\mu}} \left(\sqrt{y}\right)^{\mu-2} e^{-\frac{\sqrt{y}}{\Delta}}, \tag{6.30}$$

$$F_Y(y) = \frac{1}{\Gamma(\mu)} \Upsilon\left(\mu, \frac{\sqrt{y}}{\Delta}\right), \tag{6.31}$$

respectively, where $\mu = \frac{\delta^2}{\sigma^2}$ represents the shape parameter and $\Delta = \frac{\sigma^2}{\delta}$ represents the scale parameter. Further, let $Z \triangleq \left(\sum_{r=1}^{R} |h_{\mathrm{rf}}|\right)^2$ for an optimal phase shift. The PDF and CDF of Z are given as

$$f_Z(z) = \left(\frac{m_{\mathrm{rf}}}{\Omega_{\mathrm{rf}}}\right)^{m_{\mathrm{rf}}R} \frac{1}{\Gamma(m_{\mathrm{rf}}R)} z^{m_{\mathrm{rf}}R-1} e^{-\frac{m_{\mathrm{rf}}}{\Omega_{\mathrm{rf}}} z}, \tag{6.32}$$

$$F_Z(z) = \frac{1}{\Gamma(m_{\mathrm{rf}}R)} \Upsilon\left(m_{\mathrm{rf}}R, \frac{m_{\mathrm{rf}}}{\Omega_{\mathrm{rf}}} z\right), \tag{6.33}$$

respectively. The term $\mathcal{I}_1$ in (6.26) can be evaluated as

$$\begin{aligned} \mathcal{I}_1 &= \Pr\left[Y < \frac{\lambda \Lambda_{\mathrm{thf}} Z}{\alpha^2 \beta_{\mathrm{f}} \zeta_{\mathrm{f}}}\right] \\ &= \int_0^{\infty} f_Z(z) F_Y\left(\frac{\lambda \Lambda_{\mathrm{thf}} z}{\alpha^2 \beta_{\mathrm{f}} \zeta_{\mathrm{f}}}\right) dz. \end{aligned} \tag{6.34}$$

By substituting the corresponding PDF and CDF in (6.34) and representing $e^{-\sqrt{cz}} = \frac{1}{\sqrt{\pi}} \mathrm{G}_{0,2}^{2,0}\left[\frac{cz}{4}\Big|_{0,\frac{1}{2}}\right]$ [202, Eq. (01.03.26.0008.01)], and by using [202, Eq. (07.33.21.0088.01)], we obtain

$$\mathcal{I}_1 = 1 - \sum_{p=0}^{\mu-1} \frac{1}{p!\Gamma(m_{\mathrm{rf}}R)} \left(\frac{m_{\mathrm{rf}}}{\Omega_{\mathrm{rf}}}\right)^{-\frac{p}{2}} \frac{1}{\sqrt{\pi}} \left(\frac{1}{\Delta}\sqrt{\frac{\lambda \Lambda_{\mathrm{thf}}}{\alpha^2 \zeta_{\mathrm{f}} \beta_{\mathrm{f}}}}\right)^{p} \mathrm{G}_{1,2}^{2,1}\left[\alpha_1 \Lambda_{\mathrm{thf}}\Big|_{0,\frac{1}{2}}^{1-\frac{p}{2}-m_{\mathrm{rf}}R}\right]. \tag{6.35}$$

Further, the term $\mathcal{I}_2$ can be evaluated as

$$\begin{aligned} \mathcal{I}_2 &= \Pr\left[Y < \frac{\Lambda_{\mathrm{thf}}}{\alpha^2 \beta_{\mathrm{f}} \zeta_{\mathrm{f}}}\right] \\ &= F_Y\left(\frac{\Lambda_{\mathrm{thf}}}{\alpha^2 \beta_{\mathrm{f}} \zeta_{\mathrm{f}}}\right) \\ &= \frac{1}{\Gamma(\mu)} \Upsilon\left(\mu, \frac{1}{\Delta}\sqrt{\frac{\Lambda_{\mathrm{thf}}}{\alpha^2 \beta_{\mathrm{f}} \zeta_{\mathrm{f}}}}\right). \end{aligned} \tag{6.36}$$

By employing [197, Eq. (8.352.6)] and [202, Eq. (01.03.26.0008.01)], we obtain

$$\mathcal{I}_2 = 1 - \sum_{q=0}^{\mu-1} \frac{1}{q!\sqrt{\pi}} \left(\frac{1}{\Delta} \sqrt{\frac{\Lambda_{\text{thf}}}{\alpha^2 \zeta_{\text{f}} \beta_{\text{f}}}} \right)^q \mathrm{G}_{0,2}^{2,0} \left[\alpha_2 \Lambda_{\text{thf}} \Big|_{0,\frac{1}{2}} \right]. \tag{6.37}$$

By substituting (6.35) and (6.37) in (6.26), we obtain

$$\mathbb{P}_{\text{f}}^{\text{LB}} = 1 - \sum_{p=0}^{\mu-1} \sum_{q=0}^{\mu-1} \Psi_1 \Lambda_{\text{thf}}^{\frac{p+q}{2}} \mathrm{G}_{0,2}^{2,0} \left[\alpha_2 \Lambda_{\text{thf}} \Big|_{0,\frac{1}{2}} \right] \mathrm{G}_{1,2}^{2,1} \left[\alpha_1 \Lambda_{\text{thf}} \Big|_{0,\frac{1}{2}}^{1-\frac{p}{2}-m_{\text{rf}}R} \right], \tag{6.38}$$

where

$$\Psi_1 = \frac{\lambda^{\frac{p}{2}}}{p!q!\pi\Gamma\left(m_{\text{rf}}R\right)} \left(\frac{m_{\text{rf}}}{\Omega_{\text{rf}}} \right)^{-\frac{p}{2}} \left(\frac{1}{\Delta} \sqrt{\frac{1}{\zeta_{\text{f}} \beta_{\text{f}} \alpha^2}} \right)^{p+q}. \tag{6.39}$$

Further, by using a high SNR approximation $\Upsilon(a, y) \underset{y \to 0}{\approx} (\frac{y^a}{a})$ [211] in (6.34) and (6.36), and [197, Eq. (3.381.4)], the asymptotic OP expression for the FU is obtained as

$$\mathbb{P}_{\text{f}}^{\text{LB},\infty} \approx \mathcal{A}_1 \zeta_{\text{f}}^{-\frac{\mu}{2}} - \mathcal{A}_2 \zeta_{\text{f}}^{-\mu}, \tag{6.40}$$

where

$$\mathcal{A}_1 = \left(\frac{m_{\text{rf}}}{\Omega_{\text{rf}}} \right)^{-\frac{\mu}{2}} \frac{\Gamma\left(m_{\text{rf}}R + \frac{\mu}{2}\right)}{\mu\Gamma(\mu)\Gamma\left(m_{\text{rf}}R\right)} \left(\frac{1}{\Delta} \sqrt{\frac{\lambda \Lambda_{\text{thf}}}{\alpha^2 \beta_{\text{f}}}} \right)^{\mu} + \frac{1}{\mu\Gamma(\mu)} \left(\frac{1}{\Delta} \sqrt{\frac{\Lambda_{\text{thf}}}{\alpha^2 \beta_{\text{f}}}} \right)^{\mu}, \tag{6.41}$$

$$\mathcal{A}_2 = \left(\frac{m_{\text{rf}}}{\Omega_{\text{rf}}} \right)^{-\frac{\mu}{2}} \frac{\Gamma\left(m_{\text{rf}}R + \frac{\mu}{2}\right)}{\mu\Gamma(\mu)\Gamma\left(m_{\text{rf}}R\right)} \left(\frac{1}{\Delta} \sqrt{\frac{\lambda \Lambda_{\text{thf}}}{\alpha^2 \beta_{\text{f}}}} \right)^{\mu} \frac{1}{\mu\Gamma(\mu)} \left(\frac{1}{\Delta} \sqrt{\frac{\Lambda_{\text{thf}}}{\alpha^2 \beta_{\text{f}}}} \right)^{\mu}. \tag{6.42}$$

From (6.40), the diversity order is obtained as $\frac{\mu}{2}$, which depends on the number of REs and fading severity parameters of the BS-RIS and RIS-FU links.

6.3.2 Ergodic Capacity

Ergodic capacity for the FU is expressed as [188]

$$\mathbb{C}_{\text{f}} = \mathbb{E}\left\{ \log_2\left(1 + \Lambda_{\text{f}}\right) \right\} = \int_0^\infty \log_2\left(1 + \gamma\right) f_{\Lambda_{\text{f}}^{\text{UB}}}(\gamma) d\gamma. \tag{6.43}$$

Further, (6.43) can be simplified as [200, 201]

$$\mathbb{C}_{\mathrm{f}} = \frac{1}{\ln(2)} \int_0^{\infty} \frac{1}{1+\gamma} \left\{1 - \mathbb{P}_{\mathrm{f}}^{\mathrm{LB}}(\gamma)\right\} d\gamma. \tag{6.44}$$

By substituting (6.38) in (6.44) and representing $\frac{1}{1+\gamma} = \mathrm{G}_{1,1}^{1,1}\left[\gamma \middle|_{0}^{0}\right]$, ergodic capacity for the FU can be given as

$$\mathbb{C}_{\mathrm{f}} = \sum_{p=0}^{\mu-1}\sum_{q=0}^{\mu-1} \frac{\Psi_1}{\ln(2)} \int_0^{\infty} \gamma^{\frac{p+q}{2}} \mathrm{G}_{1,1}^{1,1}\left[\gamma \middle|_{0}^{0}\right] \mathrm{G}_{0,2}^{2,0}\left[\alpha_2\gamma \middle|_{0,\frac{1}{2}}\right] \mathrm{G}_{1,2}^{2,1}\left[\alpha_1\gamma \middle|_{0,\frac{1}{2}}^{1-\frac{p}{2}-m_{\mathrm{rf}}R}\right] d\gamma. \tag{6.45}$$

The mathematical assessment of (6.45) poses a challenge due to its complexity. Substituting $\gamma = \frac{1+x}{1-x}$, (6.45) can be simplified as

$$\begin{aligned}\mathbb{C}_{\mathrm{f}} = &\sum_{p=0}^{\mu-1}\sum_{q=0}^{\mu-1} \frac{2\Psi_1}{\ln(2)} \int_{-1}^{+1} \left(\frac{1+x}{1-x}\right)^{\frac{p+q}{2}} \frac{1}{(1-x)^2} \mathrm{G}_{1,1}^{1,1}\left[\frac{1+x}{1-x}\middle|_{0}^{0}\right] \\ &\times \mathrm{G}_{0,2}^{2,0}\left[\alpha_2\left(\frac{1+x}{1-x}\right)\middle|_{0,\frac{1}{2}}\right] \mathrm{G}_{1,2}^{2,1}\left[\alpha_1\left(\frac{1+x}{1-x}\right)\middle|_{0,\frac{1}{2}}^{1-\frac{p}{2}-m_{\mathrm{rf}}R}\right] dx. \end{aligned} \tag{6.46}$$

Now, by employing the GCQ approximation method [201] in (6.46), ergodic capacity is obtained as

$$\mathbb{C}_{\mathrm{f}} = \sum_{p=0}^{\mu-1}\sum_{q=0}^{\mu-1}\sum_{i=1}^{H} \frac{\Psi_1}{\ln(2)} \frac{2\pi\sqrt{1-x_i^2}}{H\left(1-x_i^2\right)^2} \left(\frac{1+x_i}{1-x_i}\right)^{\frac{p+q}{2}} \mathrm{G}_{1,1}^{1,1}\left[\left(\frac{1+x_i}{1-x_i}\right)\middle|_{0}^{0}\right] \mathcal{F}(x_i), \tag{6.47}$$

where $x_i = \cos\left(\frac{(2i-1)\pi}{2H}\right)$, H is a complexity-accuracy trade-off parameter [201], and

$$\mathcal{F}(x_i) = \mathrm{G}_{0,2}^{2,0}\left[\alpha_2\left(\frac{1+x_i}{1-x_i}\right)\middle|_{0,\frac{1}{2}}\right] \mathrm{G}_{1,2}^{2,1}\left[\alpha_1\left(\frac{1+x_i}{1-x_i}\right)\middle|_{0,\frac{1}{2}}^{1-\frac{p}{2}-m_{\mathrm{rf}}R}\right]. \tag{6.48}$$

6.3.3 Average Symbol Error Rate

The generalized ASER expression for any modulation scheme is given as [203, 204]

$$\mathbb{P}_{\mathrm{e,f}} = -\int_0^\infty \mathcal{P}'_s(e|\gamma)\,\mathbb{P}_{\mathrm{f}}^{\mathrm{LB}}(\gamma)\,d\gamma. \tag{6.49}$$

Rectangular QAM

By substituting the first-order derivative of conditional SEP for rectangular QAM (6.16) and the CDF of the FU (6.38) in (6.49), the ASER for rectangular QAM is obtained by using the GCQ approximation which is given as

$$\begin{aligned}
\mathbb{P}_{\mathrm{e,f}}^{\mathrm{RQAM}} =& -\frac{c_1(c_2-1)c_3}{\sqrt{2\pi}}\left(\frac{c_3^2}{2}\right)^{-\frac{1}{2}}\Gamma\left(\frac{1}{2}\right) - \frac{(c_1-1)c_2c_4}{\sqrt{2\pi}}\left(\frac{c_4^2}{2}\right)^{-\frac{1}{2}}\Gamma\left(\frac{1}{2}\right)\\
&+\sum_{t=0}^{\infty}\frac{(1)_t}{\left(\frac{3}{2}\right)_t t!}\frac{c_1c_2c_3c_4}{\pi}\left\{\left(\frac{c_3^2}{2}\right)^t+\left(\frac{c_4^2}{2}\right)^t\right\}\left(\frac{c_3^2+c_4^2}{2}\right)^{-(t+1)}\Gamma(t+1)\\
&+\sum_{p=0}^{\mu-1}\sum_{q=0}^{\mu-1}\sum_{i=1}^{H}\Psi_1\frac{c_1(c_2-1)c_3}{\sqrt{2\pi}}\frac{2\pi\sqrt{1-x_i^2}}{H\left(1-x_i^2\right)^2}\left(\frac{1+x_i}{1-x_i}\right)^{\frac{p+q-1}{2}}e^{-\frac{c_3^2}{2}\left(\frac{1+x_i}{1-x_i}\right)}\mathcal{F}(x_i)\\
&+\sum_{p=0}^{\mu-1}\sum_{q=0}^{\mu-1}\sum_{i=1}^{H}\Psi_1\frac{(c_1-1)c_2c_4}{\sqrt{2\pi}}\frac{2\pi\sqrt{1-x_i^2}}{H\left(1-x_i^2\right)^2}\left(\frac{1+x_i}{1-x_i}\right)^{\frac{p+q-1}{2}}e^{-\frac{c_4^2}{2}\left(\frac{1+x_i}{1-x_i}\right)}\mathcal{F}(x_i)\\
&-\sum_{p=0}^{\mu-1}\sum_{q=0}^{\mu-1}\sum_{t=0}^{\infty}\sum_{i=1}^{H}\Psi_1\frac{(1)_t}{\left(\frac{3}{2}\right)_t t!}\frac{c_1c_2c_3c_4}{\pi}\left\{\left(\frac{c_3^2}{2}\right)^t+\left(\frac{c_4^2}{2}\right)^t\right\}\frac{2\pi\sqrt{1-x_i^2}}{H\left(1-x_i^2\right)^2}\\
&\times\left(\frac{1+x_i}{1-x_i}\right)^{\frac{p+q}{2}+t}e^{-\frac{c_3^2+c_4^2}{2}\left(\frac{1+x_i}{1-x_i}\right)}\mathcal{F}(x_i).
\end{aligned}\tag{6.50}$$

Cross QAM

Substituting the first-order derivative of conditional SEP for cross QAM (6.19) and the CDF of the FU (6.38) in (6.49), and by using the GCQ approximation, the ASER expression is given as

$$
\begin{aligned}
\mathbb{P}_{\mathrm{e,f}}^{\mathrm{XQAM}} =& -\frac{1}{2\sqrt{\pi k_2}}\left(-k_1+\frac{12}{M_{\mathrm{a}}N_{\mathrm{a}}}+k_4\right)\left(\frac{1}{k_2}\right)^{-\frac{1}{2}}\Gamma\left(\frac{1}{2}\right)+\frac{k_5}{\sqrt{\pi k_2}}\left(\frac{k_3^2}{k_2}\right)^{-\frac{1}{2}}\Gamma\left(\frac{1}{2}\right) \\
&+\sum_{u=1}^{\frac{k_3}{2}-1}\sum_{t=0}^{\infty}\frac{(1)_t}{\left(\frac{3}{2}\right)_t t!}\frac{16u}{\pi k_2 M_a N_a}\left\{\left(\frac{1}{k_2}\right)^t+\left(\frac{4u^2}{k_2}\right)^t\right\}(k_6)^{-(t+1)}\Gamma(t+1) \\
&+\sum_{t=0}^{\infty}\frac{(1)_t}{\left(\frac{3}{2}\right)_t t!}\frac{2k_5}{\pi k_2}\left\{\left(\frac{1}{k_2}\right)^t+\left(\frac{k_3}{k_2}\right)^t\right\}\left(\frac{1+k_3^2}{k_2}\right)^{-(t+1)}\Gamma(t+1) \\
&+\sum_{t=0}^{\infty}\frac{(1)_t}{\left(\frac{3}{2}\right)_t t!}\frac{k_4}{\pi k_2}\left(\frac{1}{k_2}\right)^t\left(\frac{2}{k_2}\right)^{-(t+1)}\Gamma(t+1)+\sum_{p=0}^{\mu-1}\sum_{q=0}^{\mu-1}\sum_{i=1}^{H}\Psi_1\frac{1}{2\sqrt{\pi k_2}} \\
&\times\left(-k_1+\frac{12}{M_{\mathrm{a}}N_{\mathrm{a}}}+k_4\right)\frac{2\pi\sqrt{1-x_i^2}}{H\left(1-x_i^2\right)^2}\left(\frac{1+x_i}{1-x_i}\right)^{\frac{p+q-1}{2}}e^{-\frac{1}{k_2}\left(\frac{1+x_i}{1-x_i}\right)}\mathcal{F}(x_i) \\
&-\sum_{p=0}^{\mu-1}\sum_{q=0}^{\mu-1}\sum_{i=1}^{H}\Psi_1\frac{k_5}{\sqrt{\pi k_2}}\frac{2\pi\sqrt{1-x_i^2}}{H\left(1-x_i^2\right)^2}\left(\frac{1+x_i}{1-x_i}\right)^{\frac{p+q-1}{2}}e^{-\frac{k_3^2}{k_2}\left(\frac{1+x_i}{1-x_i}\right)}\mathcal{F}(x_i) \\
&-\sum_{p=0}^{\mu-1}\sum_{q=0}^{\mu-1}\sum_{u=1}^{\frac{k_3}{2}-1}\sum_{t=0}^{\infty}\sum_{i=1}^{H}\Psi_1\frac{(1)_t}{\left(\frac{3}{2}\right)_t t!}\frac{16u}{\pi k_2 M_a N_a}\left\{\left(\frac{1}{k_2}\right)^t+\left(\frac{4u^2}{k_2}\right)^t\right\}\frac{2\pi\sqrt{1-x_i^2}}{H\left(1-x_i^2\right)^2} \\
&\times\left(\frac{1+x_i}{1-x_i}\right)^{\frac{p+q+2t}{2}}e^{-k_6\left(\frac{1+x_i}{1-x_i}\right)}\mathcal{F}(x_i)-\sum_{p=0}^{\mu-1}\sum_{q=0}^{\mu-1}\sum_{t=0}^{\infty}\sum_{i=1}^{H}\Psi_1\frac{(1)_t}{\left(\frac{3}{2}\right)_t t!}\frac{2k_5}{\pi k_2} \\
&\times\left\{\left(\frac{1}{k_2}\right)^t+\left(\frac{k_3}{k_2}\right)^t\right\}\frac{2\pi\sqrt{1-x_i^2}}{H\left(1-x_i^2\right)^2}\left(\frac{1+x_i}{1-x_i}\right)^{\frac{p+q+2t}{2}}e^{-\left(\frac{1+k_3^2}{2}\right)\left(\frac{1+x_i}{1-x_i}\right)}\mathcal{F}(x_i) \\
&-\sum_{p=0}^{\mu-1}\sum_{q=0}^{\mu-1}\sum_{t=0}^{\infty}\sum_{i=1}^{H}\Psi_1\frac{(1)_t}{\left(\frac{3}{2}\right)_t t!}\frac{k_4}{\pi k_2}\left(\frac{1}{k_2}\right)^t\frac{2\pi\sqrt{1-x_i^2}}{H\left(1-x_i^2\right)^2}\left(\frac{1+x_i}{1-x_i}\right)^{\frac{p+q+2t}{2}} \\
&\times e^{-\frac{2}{k_2}\left(\frac{1+x_i}{1-x_i}\right)}\mathcal{F}(x_i).
\end{aligned}
\tag{6.51}
$$

Hexagonal QAM

By substituting the first-order derivative of conditional SEP for hexagonal QAM (6.22) and the CDF of the FU (6.38) in (6.49), the ASER for hexagonal QAM is obtained by using the GCQ approximation which is given as

$$
\begin{aligned}
\mathbb{P}_{\text{e,f}}^{\text{HQAM}} =& -\frac{1}{2}\sqrt{\frac{g_3}{2\pi}}\left(g_2-g_1\right)\left(\frac{g_3}{2}\right)^{-\frac{1}{2}}\Gamma\left(\frac{1}{2}\right)+\frac{g_2}{3}\sqrt{\frac{g_3}{3\pi}}\left(\frac{g_3}{3}\right)^{-\frac{1}{2}}\Gamma\left(\frac{1}{2}\right)\\
&-\frac{g_2}{2}\sqrt{\frac{g_3}{6\pi}}\left(\frac{g_3}{6}\right)^{-\frac{1}{2}}\Gamma\left(\frac{1}{2}\right)-\sum_{t=0}^{\infty}\frac{(1)_t}{\left(\frac{3}{2}\right)_t t!}\frac{2g_2g_3}{9\pi}\left(\frac{g_3}{3}\right)^t\left(\frac{2g_3}{3}\right)^{-(t+1)}\\
&\times\Gamma\left(t+1\right)+\sum_{t=0}^{\infty}\frac{(1)_t}{\left(\frac{3}{2}\right)_t t!}\frac{g_2g_3}{2\sqrt{3}\pi}\left\{\left(\frac{g_3}{2}\right)^t+\left(\frac{g_3}{6}\right)^t\right\}\left(\frac{2g_3}{3}\right)^{-(t+1)}\Gamma\left(t+1\right)\\
&+\sum_{p=0}^{\mu-1}\sum_{q=0}^{\mu-1}\sum_{i=1}^{H}\Psi_1\frac{1}{2}\sqrt{\frac{g_3}{2\pi}}\left(g_2-g_1\right)\frac{2\pi\sqrt{1-x_i^2}}{H\left(1-x_i^2\right)^2}\left(\frac{1+x_i}{1-x_i}\right)^{\frac{p+q-1}{2}}e^{-\frac{g_3}{2}\left(\frac{1+x_i}{1-x_i}\right)}\mathcal{F}\left(x_i\right)\\
&-\sum_{p=0}^{\mu-1}\sum_{q=0}^{\mu-1}\sum_{i=1}^{H}\Psi_1\frac{g_2}{3}\sqrt{\frac{g_3}{3\pi}}\frac{2\pi\sqrt{1-x_i^2}}{H\left(1-x_i^2\right)^2}\left(\frac{1+x_i}{1-x_i}\right)^{\frac{p+q-1}{2}}e^{-\frac{g_3}{3}\left(\frac{1+x_i}{1-x_i}\right)}\mathcal{F}\left(x_i\right)\\
&+\sum_{p=0}^{\mu-1}\sum_{q=0}^{\mu-1}\sum_{i=1}^{H}\Psi_1\frac{g_2}{2}\sqrt{\frac{g_3}{6\pi}}\frac{2\pi\sqrt{1-x_i^2}}{H\left(1-x_i^2\right)^2}\left(\frac{1+x_i}{1-x_i}\right)^{\frac{p+q-1}{2}}e^{-\frac{g_3}{6}\left(\frac{1+x_i}{1-x_i}\right)}\mathcal{F}\left(x_i\right)\\
&+\sum_{p=0}^{\mu-1}\sum_{q=0}^{\mu-1}\sum_{t=0}^{\infty}\sum_{i=1}^{H}\Psi_1\frac{(1)_t}{\left(\frac{3}{2}\right)_t t!}\frac{2g_2g_3}{9\pi}\left(\frac{g_3}{3}\right)^t\frac{2\pi\sqrt{1-x_i^2}}{H\left(1-x_i^2\right)^2}\left(\frac{1+x_i}{1-x_i}\right)^{\frac{p+q+2t}{2}}\\
&\times e^{-\frac{2g_3}{3}\left(\frac{1+x_i}{1-x_i}\right)}\mathcal{F}\left(x_i\right)-\sum_{p=0}^{\mu-1}\sum_{q=0}^{\mu-1}\sum_{t=0}^{\infty}\sum_{i=1}^{H}\Psi_1\frac{(1)_t}{\left(\frac{3}{2}\right)_t t!}\frac{g_2g_3}{2\sqrt{3}\pi}\\
&\times\left\{\left(\frac{g_3}{2}\right)^t+\left(\frac{g_3}{6}\right)^t\right\}\frac{2\pi\sqrt{1-x_i^2}}{H\left(1-x_i^2\right)^2}\left(\frac{1+x_i}{1-x_i}\right)^{\frac{p+q+2t}{2}}e^{-\frac{2g_3}{3}\left(\frac{1+x_i}{1-x_i}\right)}\mathcal{F}\left(x_i\right).
\end{aligned}
\tag{6.52}
$$

6.4 Numerical and Simulation Results

This section presents the numerical results for evaluating the performance of the considered network. The accuracy of the derived closed-form expressions for OP, asymptotic OP, ergodic capacity, and ASER are validated using Monte Carlo simulations. The distances between the BS and NU, BS and active RIS, and active RIS and FU are set to 10, 40, and 10 m, respectively [188]. The parameters $\alpha = 2$ dB, $\beta_{\text{n}} = 0.3$, $\beta_{\text{f}} = 0.7$, and $\sigma_{\text{n}}^2 = \sigma_{\text{f}}^2 = \sigma_{\text{r}}^2 = -50$ dBm are set unless specified [184, 212]. The fading severity parameters m_{sr} and m_{rf} are denoted as $\{m_{\text{sr}}, m_{\text{rf}}\}$ in the figures, and consider $\mathcal{R}_{\text{n}} = \mathcal{R}_{\text{f}} = \text{r}_{\text{th}}$ for simplicity. The numerical and simulation results align perfectly, validating the accuracy of the derived closed-form expressions. Note that OMA is treated as a special case of NOMA. For a fair comparison, both the NU and FU share half of the resource block, with $\Lambda_{\text{thf}} = 2^{2\mathcal{R}_{\text{f}}} - 1$ and $\Lambda_{\text{thn}} = 2^{2\mathcal{R}_{\text{n}}} - 1$. Furthermore, the system performance with a passive RIS is derived by setting $\alpha_{\text{r}} = 1$ and $\sigma_{\text{r}}^2 = 0$.

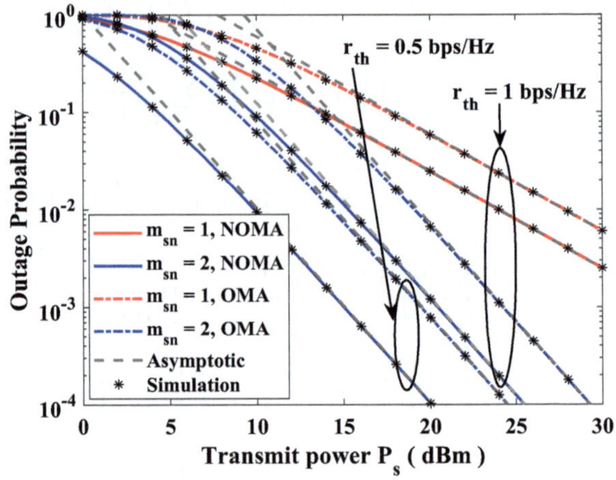

Fig. 6.2 Outage probability versus transmit power of NU with $\sigma_n^2 = \sigma_f^2 = -30$ dBm

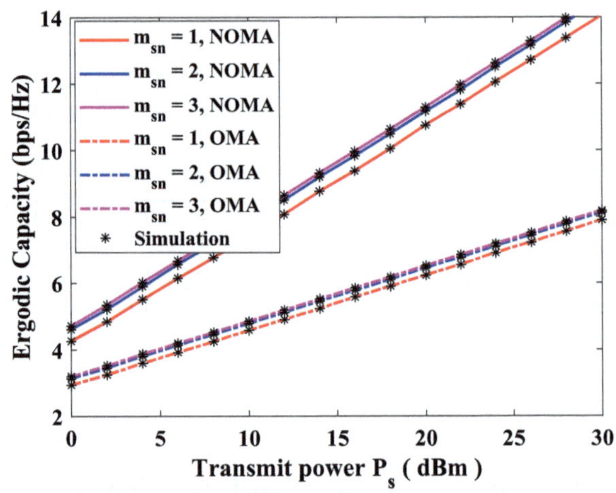

Fig. 6.3 Ergodic capacity versus transmit power of NU

6.4.1 Near User

Figure 6.2 depicts NU OP versus transmit power curves. An improvement in OP is observed as the severity parameter m_{sn} increases, as this indicates better channel conditions for signal transmission. To attain an OP of 10^{-2}, approximately 8.83 dB gain is achieved as the severity parameter m_{sn} increases from 1 to 2, with $r_{th} = 1$ bps/Hz. Additionally, it is observed that outage performance improves as r_{th} decreases, since a lower target rate results in a relatively lower target SNR. This insight is valuable for designing a communication setup with a specific target rate. Moreover, it is evident that NOMA outperforms OMA. For $m_{sn} = 1$ and 2, the considered NOMA network respectively saves approximately 3.82 and 3.80 dB in transmit power with respect to OMA to attain an OP of 10^{-2} with $r_{th} = 1$ bps/Hz.

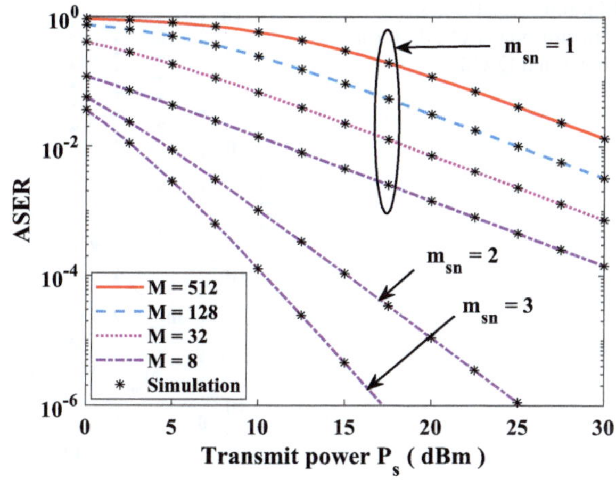

Fig. 6.4 ASER versus transmit power of NU for a rectangular QAM

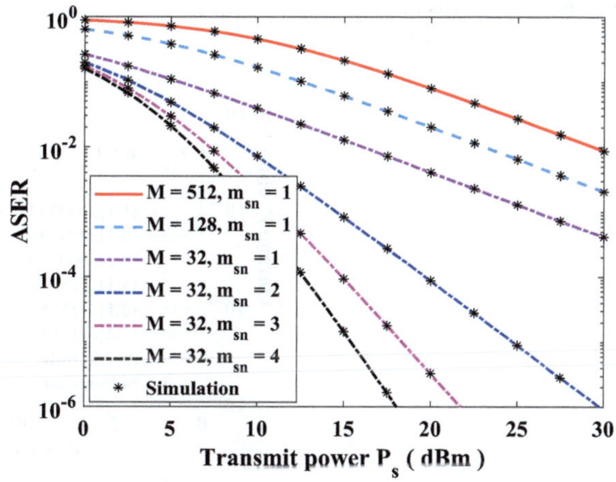

Fig. 6.5 ASER versus transmit power of NU for a cross QAM

Figure 6.3 illustrates the ergodic capacity of the NU as a function of transmit power for different severity parameters. It can be observed that the ergodic capacity increases with both transmit power P_s and the severity parameter m_{sn}. To achieve an ergodic capacity of 6 bps/Hz, approximately 1.26 and 1.72 dB power saving gains are achieved as m_{sn} increases from 1 to 2 and 3, respectively. Furthermore, to achieve an ergodic capacity of 6 bps/Hz, the considered NOMA requires approximately 13.20, 12.96, and 12.82 dB less power than OMA for $m_{sn} = 1$, 2, and 3, respectively.

Figures 6.4, 6.5, and 6.6 depict the ASER versus transmit power curves of the NU under rectangular QAM, cross QAM, and hexagonal QAM schemes, respectively. The ASER performance improves with m_{sn} due to enhanced channel conditions for signal transmission. Meanwhile, the ASER performance degrades as the constellation size M increases. From Fig. 6.4, it is clear that achieving an ASER of 10^{-2} results in power savings of approximately 6.45 and 13.56 dB when M decreases from 128 to

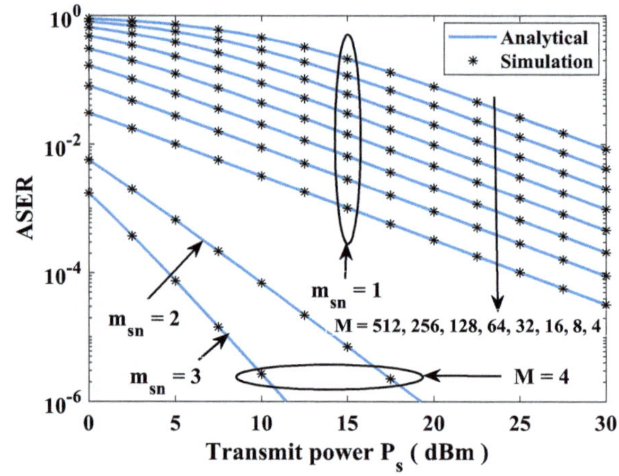

Fig. 6.6 ASER versus transmit power of NU for a hexagonal QAM

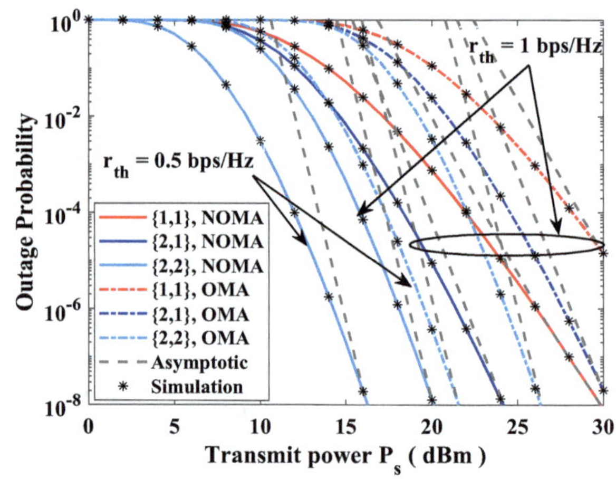

Fig. 6.7 OP versus transmit power of FU with $R = 8$

32 and 8, respectively, with $m_{\text{sn}} = 1$. Further, from Fig. 6.5, it is noticed that when M decreases from 512 to 128 and 32, approximately 6.21 and 13.32 dB power saving gains are respectively achieved to attain an ASER of 10^{-2} with $m_{\text{sn}} = 1$. Furthermore, it can be observed from Fig. 6.6 that to achieve an ASER of 10^{-2}, approximately 3.05, 6.19, 9.43, 12.97, 16.04, 19.74, and 24.16 dB power saving gains are achieved when M decreases from 512 to 256, 128, 64, 32, 16, 8, and 4, respectively, with $m_{\text{sn}} = 1$.

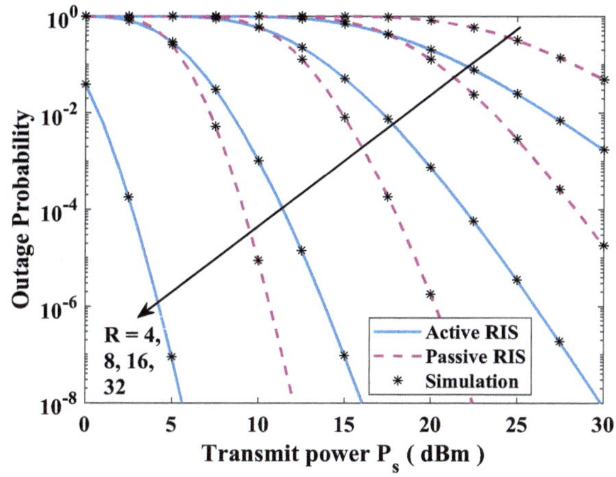

Fig. 6.8 OP versus transmit power of FU with severity parameter {1, 1} and $r_{th} = 1$ bps/Hz

6.4.2 *Far User*

Figure 6.7 depicts FU OP versus transmit power for $R = 8$. Additionally, the asymptotic OP approaches the OP curves in the high transmit SNR regimes. It can be observed that the OP decreases with the severity parameters $\{m_{sr}, m_{rf}\}$. To achieve an OP of 10^{-2}, approximately 2.45 and 4.20 dB power saving gains are achieved as $\{m_{sr}, m_{rf}\}$ parameters transition from {1, 1} to {2, 1} and {2, 2}, respectively, with $r_{th} = 1$ bps/Hz. Furthermore, it is observed that OP performance improves as the target rate decreases, as a lower target rate requires a relatively lower SNR. Additionally, a significant improvement in OP for NOMA compared to OMA is observed. For $\{m_{sr}, m_{rf}\}$ parameters {1, 1}, {2, 1}, and {2, 2}, NOMA achieves respectively 6.16, 6.24, and 6.19 dB power saving gains compared with the OMA.

Figure 6.8 illustrates the beneficial effect of increasing the number of active REs on the FU's OP performance. To achieve an OP of 10^{-2}, approximately 9.59, 18.26, and 26.01, dB power saving gains are noticed as R increases from 4 to 8, 16, and 32, respectively. This significant improvement is attributed to the enhanced signal strength and fine-grained reflective beamforming as R increases, enabling the deployment of a communication setup that achieves the desired performance level in practice. Moreover, the system utilizing an active RIS demonstrates substantial performance gains over a passive RIS implementation, particularly in terms of transmit power savings P_s. To achieve an OP of 10^{-2}, power savings of approximately 6.41, 6.44, and 6.39 dB power are obtained for $R = 8$, 16, and 32, respectively.

Figure 6.9 illustrates the influence of the amplification gain α and the thermal noise power σ_r^2 introduced by the active REs on the FU's OP. It is noted that the OP improves with α due to the increased signal strength. To achieve an OP of 10^{-2}, approximately 1.81, 3.87, and 5.80 dB power saving gains are attained when α increases from 2 to 3, 4, and 5 dB, respectively, with $\sigma_r^2 = -50$ dBm. This indicates that the amplification gain provided by the RIS can support the practical imple-

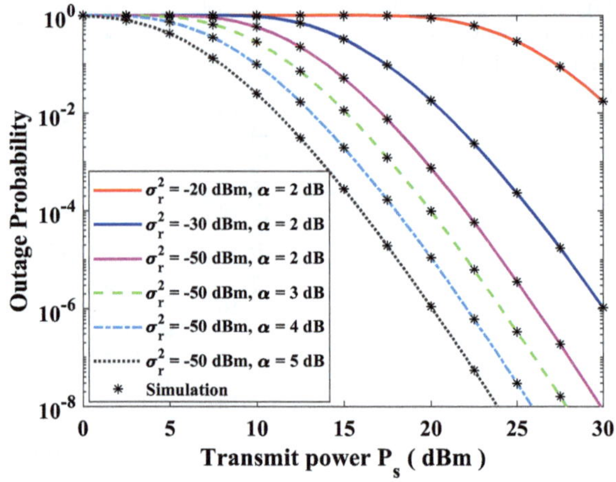

Fig. 6.9 OP versus transmit power of FU with $R = 8$ and $r_{th} = 1$ bps/Hz

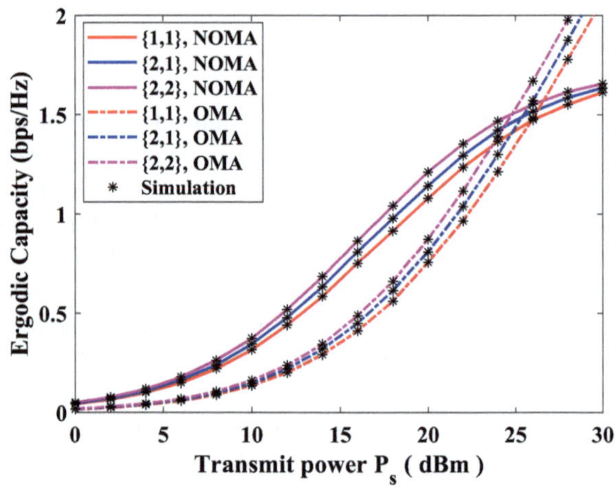

Fig. 6.10 Ergodic capacity versus transmit power of FU with $R = 4$

mentation of energy-efficient wireless communication systems. Moreover, the OP decreases as σ_r^2 is reduced. Indeed, when σ_r^2 reduces from -20 dBm to -30 and -50 dBm, respectively 9.68, and 13.09 dB power saving gains are attained to achieve an OP of 10^{-1} with $\alpha = 2$ dB.

Figure 6.10 shows the ergodic capacity of the FU as a function of transmit power. In the considered NOMA network, the ergodic capacity increases with transmit power P_s in the low transmit power regime. However, as transmit power continues to rise, NOMA performance reaches a saturation point due to increased interference experienced by the NU. Despite this, NOMA outperforms OMA in the low transmit power regime. Additionally, the ergodic capacity improves with better fading statistics.

Figure 6.11 illustrates the ergodic capacity of the FU as a function of transmit power with a severity parameter of $\{2, 2\}$, highlighting the performance improvement as the number of REs increases. Additionally, the active RIS implementation

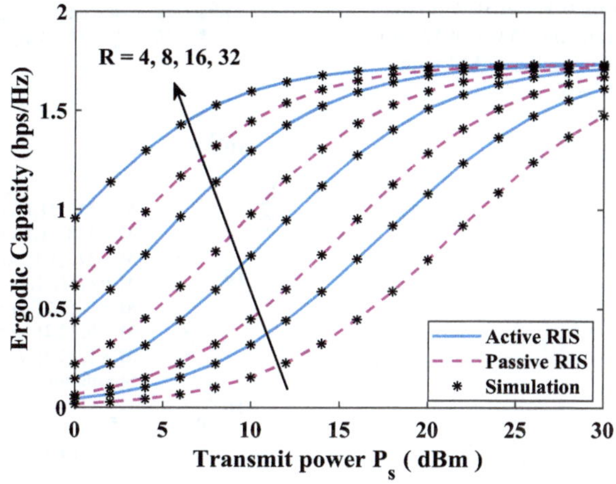

Fig. 6.11 Ergodic capacity versus transmit power of FU with severity parameter {2, 2}

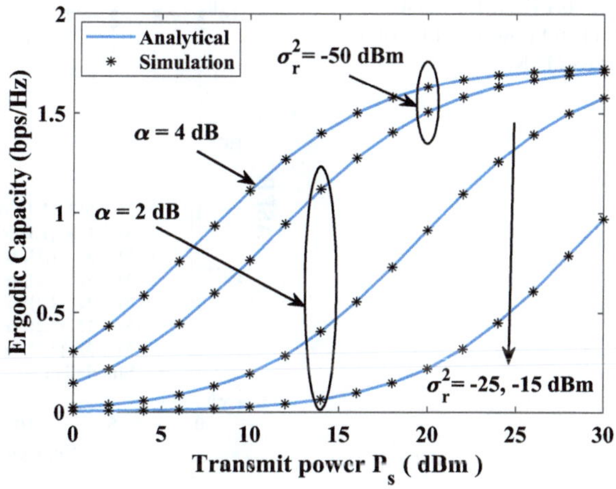

Fig. 6.12 Ergodic capacity versus transmit power of FU with severity parameters {1, 1}

outperforms the passive RIS implementation due to the amplification gain provided by the former. Indeed, to attain an ergodic capacity of 1 bps/Hz, approximately 3.91, 3.87, 3.82, and 3.64 dB power saving gains are achieved with the active RIS compared to the passive RIS implementation for $R = 4$, 8, 16, and 32, respectively.

Figure 6.12 illustrates the positive and negative effects of the amplification gain α and thermal noise power σ_r^2 introduced by the active REs on the FU's ergodic capacity performance, respectively. To achieve an ergodic capacity of 1 bps/Hz, approximately 3.85 dB power saving gain is observed as α increases from 2 to 4 dB, with $\sigma_r^2 = -50$ dBm. However, when σ_r^2 decreases from -15 to -25 and -50 dBm, approximately 9.42 and 17.72 dB respective power saving gains are observed to achieve an ergodic capacity of 0.8 bps/Hz with $\alpha = 2$ dB.

Figures 6.13, 6.14, and 6.15 illustrate the ASER performance versus transmit power for rectangular QAM, cross QAM, and hexagonal QAM schemes, respectively.

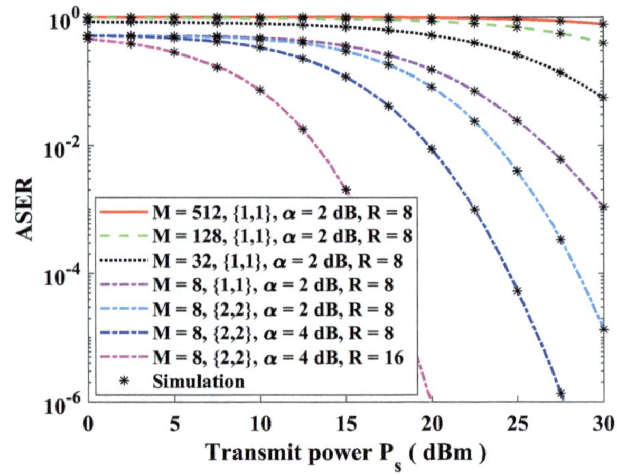

Fig. 6.13 ASER versus transmit power of FU for a rectangular QAM with $R = 8$

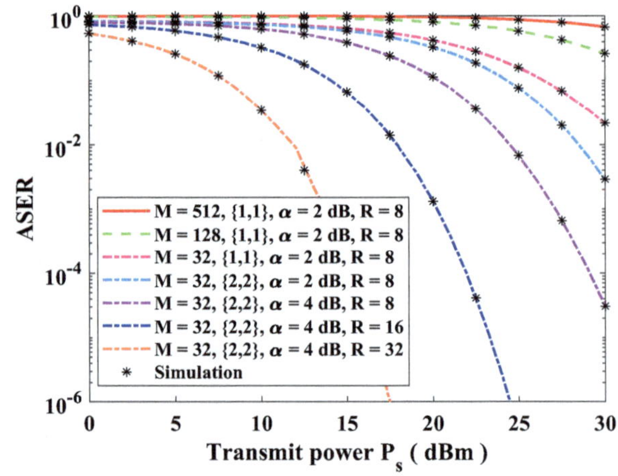

Fig. 6.14 ASER versus transmit power of FU for a cross QAM

It is observed that the ASER performance improves with $\{m_{sr}, m_{rf}\}$, amplification gain α, and number of REs R. However, the ASER performance worsens as the constellation size M increases. From Fig. 6.13, it is observed that to achieve an ASER of 10^{-2}, approximately 2.87, 3.99, and 6.56 dB power saving gains are attained when $\{m_{sr}, m_{rf}\}$ increases from $\{1, 1\}$ to $\{2, 2\}$, α increases from 2 to 4, and R increases from 8 to 16, respectively, with $M = 8$. From Fig. 6.14, it is observed that to attain an ASER of 10^{-1}, approximately 2.08, 4.01, and 6.29 dB power saving gains are achieved when $\{m_{sr}, m_{rf}\}$ increases from $\{1, 1\}$ to $\{2, 2\}$, α increases from 2 to 4, and R increases from 8 to 16, respectively, with $M = 32$. Furthermore, from Fig. 6.15, it is observed that to attain an ASER of 10^{-2}, approximately 3.02, 3.99, and 6.16 dB power saving gains are attained when $\{m_{sr}, m_{rf}\}$ increases from $\{1, 1\}$ to $\{2, 2\}$, α increases from 2 to 4, and R increases from 8 to 16, respectively, with $M = 16$.

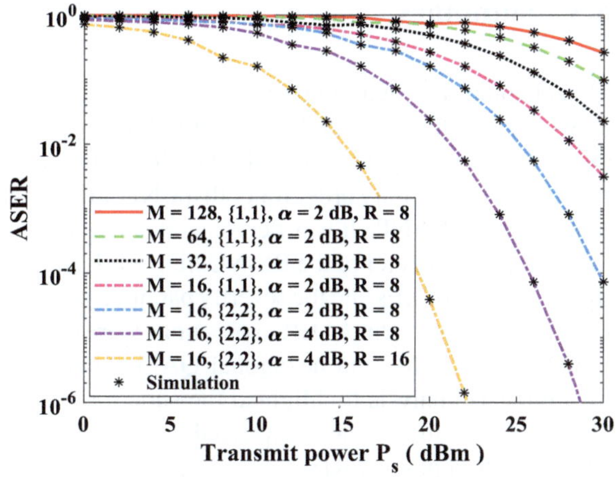

Fig. 6.15 ASER versus transmit power of FU for a hexagonal QAM

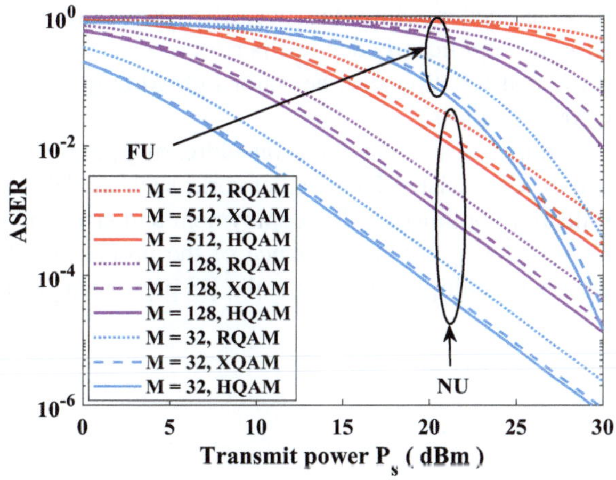

Fig. 6.16 Comparison among various QAM schemes

Figure 6.16 presents a comparison of ASER performance among rectangular QAM, cross QAM, and hexagonal QAM schemes. It is evident that cross QAM consistently achieves a notable performance improvement over rectangular QAM across all constellation orders. This is because cross QAM has lower peak and average power requirements than rectangular QAM due to its more compact design. Moreover, hexagonal QAM exhibits a significant performance gain over cross QAM, establishing itself as the optimal constellation scheme.

6.5 Summary

This chapter analyzed the performance of an active RIS-assisted downlink NOMA network operating over Nakagami-m fading channels. Closed-form expressions for OP and ergodic capacity were derived for both the NU and FU using the Meijer-G function. The diversity order for both the NU and FU was determined from the asymptotic OP at high SNR. It was demonstrated that the NU's diversity order depends on the severity parameter of the BS-NU link, while the FU's diversity order is influenced by the number of REs and the severity parameters of the BS-active RIS and active RIS-FU links. Closed-form ASER expressions were derived for general-order rectangular QAM, cross QAM, and hexagonal QAM schemes using the GCQ approximation method. The analysis highlighted the superiority of NOMA over OMA in terms of OP and demonstrated the exceptional performance of the active RIS implementation compared to a passive RIS-assisted system. Numerical results further revealed that system performance improves with increased amplification gain, a higher number of REs, and better fading statistics, whereas the thermal noise introduced by the active REs can significantly degrade performance. The ASER analysis indicated that, for a given constellation size, hexagonal QAM outperforms rectangular QAM and cross QAM schemes.

Future research can focus on addressing potential challenges in implementing an active RIS-assisted NOMA network, such as RIS reconfiguration for controllable reflection, optimizing RIS deployment and size, and channel estimation. Additionally, exploring the performance of the proposed network with multiple downlink users and multiple antennas at the BS presents a promising research direction.

Chapter 7
Conclusion and Future Work

This book provides a comprehensive exploration of advanced NOMA techniques in heterogeneous cellular networks, addressing various real-world challenges and enhancing system performance. Beginning with an introduction to NOMA, cooperative NOMA, performance metrics, and key enabling technologies such as SWIPT and RISs, the discussion lays the foundation for an in-depth analysis of these techniques.

A detailed investigation is conducted into the impact of non-linear distortions in high-power amplifiers and imperfect successive interference cancelation, demonstrating that NOMA significantly outperforms OMA even under non-ideal conditions, achieving substantial performance improvements. The analysis is further extended by incorporating the effects of imperfect channel state information, providing critical insights into its interplay with interference cancelation and its influence on spectral and energy efficiencies.

The study also explores a SWIPT-enabled cooperative NOMA framework, where cooperative communication is leveraged to enhance the performance of cell-edge users while ensuring energy-efficient operation. Analytical results highlight the advantages of SWIPT-based cooperative NOMA over conventional energy harvesting models, offering new perspectives on energy efficiency in heterogeneous networks. Additionally, the impact of impulsive noise in industrial IoT applications is examined, showcasing NOMA's robustness against such interference and emphasizing its spectral and energy efficiency gains over OMA under both Gaussian and impulsive noise conditions.

The role of RISs in NOMA-based downlink networks is also investigated, with analytical expressions derived for outage probability, ergodic capacity, and symbol error rates in advanced modulation schemes. The findings underscore the potential of RIS in optimizing wireless propagation environments to further enhance NOMA performance.

Through these extensive analyses and insights, this book establishes a strong foundation for future research on advanced NOMA techniques in heterogeneous cellular networks. The discussions and findings presented will serve as a valuable

V. Bhatia et al., *Advanced NOMA Techniques for Heterogeneous Cellular Networks*, Wireless Networks, https://doi.org/10.1007/978-3-032-09551-0_7

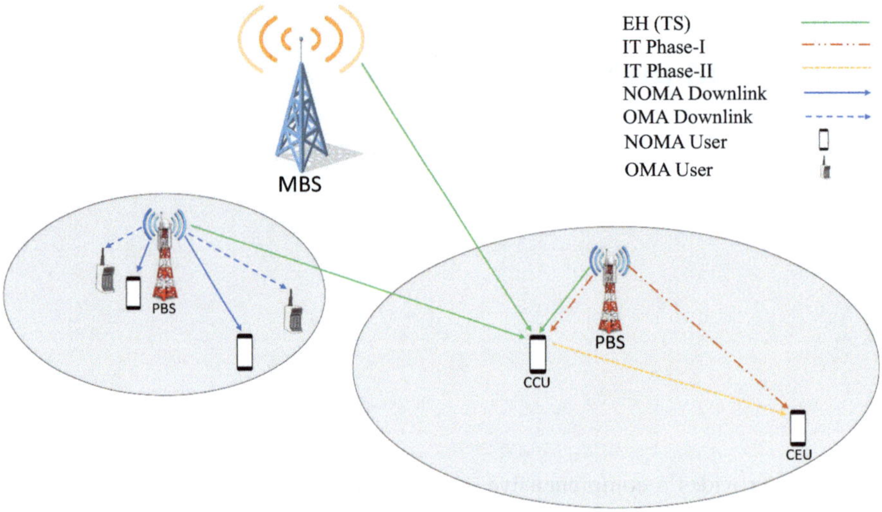

Fig. 7.1 System model of SWIPT-enabled Co-NOMA systems

resource for both academia and industry professionals, driving the development of more efficient, resilient, and scalable next-generation wireless networks. The future work on NOMA-based communication is listed below.

7.1 Enhancing SWIPT-Enabled Co-NOMA in HCNs: Addressing NL-HPA Distortions and Future Directions

Building on the insights gained in this study, future research will delve deeper into optimizing the performance of SWIPT-enabled Co-NOMA systems in HCNs under the influence of NL-HPA distortions. A primary focus will be the development of advanced interference exploitation strategies that harness constructive interference to enhance energy efficiency and overall system performance. These strategies will dynamically adapt to varying interference levels, ensuring improved energy harvesting capabilities for SWIPT users and reliable communication for end users.

To address the challenge of NL-HPA distortions, ML-based predictive models will be developed to accurately characterize and compensate for non-linear effects in real time. These models will also facilitate adaptive resource allocation strategies that optimize power and spectrum usage in multi-tier HCNs. Additionally, the implementation of advanced error correction and coding schemes will be considered to mitigate the impact of non-linear distortions on system reliability (Fig. 7.1).

This research will examine the applicability of the proposed methods to support emerging use cases, such as URLLC and industrial IoT applications. By addressing practical challenges and incorporating cutting-edge technologies, future studies aim to establish a robust foundation for deploying SWIPT-enabled Co-NOMA systems in next-generation HCNs, ensuring their scalability and efficiency in diverse real-world scenarios.

7.2 Active RIS-Aided Co-NOMA: Energy Efficiency, Resource Optimization, and Multi-RIS Cooperation

Future work will also investigate the integration of RIS to overcome limitations imposed by NL-HPA. RIS-aided Co-NOMA systems have the potential to reshape wireless propagation environments, enabling precise interference control and improved signal quality for end users. Furthermore, full-duplex and half-duplex user relaying techniques will be explored to enhance cooperation phases and reduce the performance trade-offs observed between users. Future research on active RIS-aided NOMA networks holds tremendous promise for transforming next-generation wireless communication systems by tackling key challenges and exploring new opportunities. A critical area of focus is improving energy efficiency, as active RIS, unlike its passive counterpart, requires power to amplify signals. This calls for the development of advanced algorithms that effectively balance power consumption and performance. Additionally, optimizing resource allocation is essential for managing power, spectrum, and RIS elements in dynamic environments, ensuring maximum throughput and fairness among users. Joint user scheduling and active RIS configuration also present significant research opportunities, necessitating intelligent solutions capable of adapting to varying user mobility and channel conditions. Accurate channel estimation, which becomes more complex in active RIS systems due to signal amplification, opens doors to innovative approaches leveraging machine learning and sparse signal processing techniques. Security and privacy are also critical concerns, as the amplified signals in active RIS may introduce vulnerabilities such as signal interception, making the design of secure transmission protocols and privacy-preserving mechanisms imperative. Furthermore, integrating active RIS with advanced technologies like massive MIMO, artificial intelligence, and 6G applications such as URLLC and holographic transmission can significantly enhance system capabilities. Exploring hybrid RIS configurations that combine active and passive elements is another promising direction for achieving an optimal balance between cost, power consumption, and performance. Real-world deployment studies are equally important, requiring investigations into the impact of hardware imperfections, environmental dynamics, and non-linearities, along with experimental validation of proposed designs. Active RIS also has the potential to greatly improve cell-edge user performance in NOMA networks, ensuring fairness and efficiency across the network. Finally, cooperative strategies for multi-RIS deployments, focusing on inter-RIS communication and synchronization, can further boost scalability and coverage (Fig. 7.2).

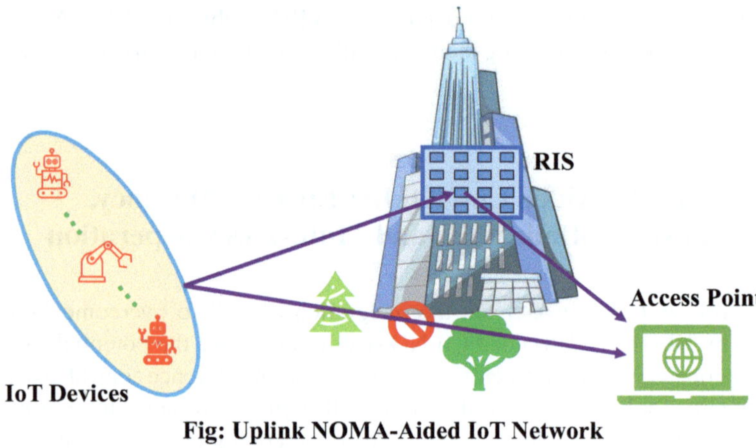

Fig. 7.2 System model of uplink NOMA systems

7.3 Exploring Active RIS-Aided NOMA with Multi-antenna BS: Challenges and Future Research Directions

Furthermore, investigating the performance of active RIS-aided NOMA networks with multiple downlink users and multiple antennas at the BS offers a highly promising avenue for advancing next-generation wireless communication systems. This configuration presents unique challenges and opportunities, as integrating active RIS with multi-antenna BS setups can enhance spatial diversity, spectral efficiency, and fairness among users. Future research could focus on designing advanced beamforming strategies that optimize signal transmission and reception while considering the amplified signal characteristics of active RIS. Moreover, the joint optimization of power allocation, user pairing, and RIS phase shifts in multi-user, multi-antenna scenarios is critical to achieving superior system performance. The interaction between active RIS and the BS's spatial multiplexing capabilities also calls for innovative interference management techniques and efficient exploitation of additional degrees of freedom. Incorporating machine learning algorithms to dynamically optimize resource allocation and RIS configurations in real-time could further improve performance in dynamic environments. Additionally, assessing the effects of practical constraints, such as channel estimation errors, hardware impairments, and user mobility, is crucial to ensuring the robustness and feasibility of proposed solutions. By addressing these challenges, future studies can fully harness the potential of active RIS-aided NOMA networks in complex multi-user, multi-antenna scenarios, contributing to the development of efficient and scalable wireless systems. By addressing these diverse challenges, future research will unlock the full potential of active RIS-assisted NOMA network to revolutionize wireless networks.

Appendix A
Proofs of Chapter 2

A.1 Proof of Proposition 2.1

We have provided a detailed step-by-step explanation for the derivation of the expression given as Eq. (2.22), based on the Gauss-Chebyshev (G-C) Quadrature approximation:

1. Starting from the expression (2.21), the CDF of the MBS tier's unordered channel gain using polar coordinates and the homogeneous PPP assumption can be expressed as follows.

$$F_{|\widetilde{h}_m|^2}(y) = 2\pi\lambda_m \int_0^{\mathcal{Y}_m} (1 - e^{-(1+r_m^{\nu_m})y}) e^{-\pi\lambda_m r_m^2} r_m dr_m, \tag{A.1}$$

where λ_m is the node density, and $\mathcal{Y}_m$ denotes the coverage range of the MBS tier. For the MBS tier, the path loss exponent is given as ν_m and the distance between the tagged BS (BS that serves the typical user) and the user is given as r_m.
2. We applied the G-C quadrature approximation technique to this expression. It is an effective approach to calculate the integral of the mathematically intractable function using G-C quadrature approximation. It is defined as:

$$\int_a^b f(x)dx \approx \frac{b-a}{2} \sum_{i=1}^{I} \omega_i \sqrt{1-\theta_i^2} f(x_i), \tag{A.2}$$

where $\omega_i = \frac{\pi}{I}, x_i = \frac{b-a}{2}\theta_i + \frac{b+a}{2}$, and $\theta_i = cos\left(\frac{2i-1}{2I}\pi\right), \forall i = 1, 2, ..., I$. The G-C quadrature approximation is commonly applied in the literature to approximate the CDF as in [114, Eq. (42)], [141, Eq. (9)], [140, Eq. (19)], and [110, Eq. (4)]. The approximation is also used in literature to approximate mathematically intractable integrals as in [213, Eq. (10)], [214, Eq. (9)], and [215, Eq. (20)].
3. On comparing (A.1) and (A.2), we substitute $a = 0$, $b = \mathcal{Y}_m$, and $f(x) = (1 - e^{-(1+x^{\nu_m})y}) e^{-\pi\lambda_m x^2} x$ in (A.2), we get

V. Bhatia et al., *Advanced NOMA Techniques for Heterogeneous Cellular Networks*, Wireless Networks, https://doi.org/10.1007/978-3-032-09551-0

$$\int_0^{\mathcal{Y}_m}\left(1-e^{-(1+x^{\nu_m})y}\right)e^{-\pi\lambda_m x^2}x\,dx \approx \frac{\mathcal{Y}_m}{2}\sum_{i=1}^{I}\omega_i\sqrt{1-\theta_i^2}\left(1-e^{-(1+x_i^{\nu_m})y}\right)e^{-\pi\lambda_m x_i^2}x_i, \tag{A.3}$$

where $x_i = \frac{\mathcal{Y}_m}{2}(\theta_i + 1)$. On substituting the value of x_i in the (A.3), we get

$$\int_0^{\mathcal{Y}_m}\left(1-e^{-(1+x^{\nu_m})y}\right)e^{-\pi\lambda_m x^2}x\,dx \approx \frac{\mathcal{Y}_m}{2}\sum_{i=1}^{I}\omega_i\sqrt{1-\theta_i^2}\left(1-e^{-\left(1+\left(\frac{\mathcal{Y}_m}{2}\theta_i+\frac{\mathcal{Y}_m}{2}\right)^{\nu_m}\right)y}\right)$$
$$\times e^{-\pi\lambda_m\left(\frac{\mathcal{Y}_m}{2}(\theta_i+1)\right)^2}\frac{\mathcal{Y}_m}{2}(\theta_i+1), \tag{A.4}$$

on substituting $b_i^m = -\omega_i\sqrt{1-\theta_i^2}\left(\frac{1}{2}(\theta_i+1)\right)e^{-\pi\lambda_m\left(\frac{1}{2}(\theta_i+1)\mathcal{Y}_m\right)^2}$, and $c_i^m = 1 + \left(\frac{\mathcal{Y}_m}{2}\theta_i + \frac{\mathcal{Y}_m}{2}\right)^{\nu_m}$ in the expression above, and considering $b_0^m = -\sum_{i=1}^{I} b_i^m$, and $c_0^m = 0$, we get

$$\begin{aligned}\int_0^{\mathcal{Y}_m}\left(1-e^{-(1+x^{\nu_m})y}\right)e^{-\pi\lambda_m x^2}x\,dx &\approx \frac{\mathcal{Y}_m^2}{2}\sum_{i=1}^{I}-b_i^m\left(1-e^{-c_i^m y}\right)\\ &\approx \frac{\mathcal{Y}_m^2}{2}\left(\sum_{i=1}^{I}-b_i^m+\sum_{i=1}^{I}b_i^m e^{-c_i^m y}\right)\\ &\approx \frac{\mathcal{Y}_m^2}{2}\left(b_0^m e^{-c_0^m y}+\sum_{i=1}^{I}b_i^m e^{-c_i^m y}\right)\\ &\approx \frac{\mathcal{Y}_m^2}{2}\sum_{i=0}^{I}b_i^m e^{-c_i^m y}\end{aligned} \tag{A.5}$$

4. On substituting expression from Eq. (A.5) in CDF Eq. (A.1), unordered CDF expression can be written as,

$$F_{|\tilde{h}_m|^2}(y) \simeq \pi\lambda_m\mathcal{Y}_m^2\sum_{\iota=0}^{I}b_\iota^m e^{-c_\iota^m y}. \tag{A.6}$$

A.2 Proof of Proposition 2.2

We will now proceed with a comprehensive derivation of equation (2.33) to further elucidate the key mathematical relation between CDF of ordered and unordered channel gain.

1. From Binomial expansion, we may write,

$$\left(1-F_{\left|\tilde{h}_{po}\right|^2}(y)\right)^j = \sum_{d=0}^{j}\binom{j}{d}(1)^{j-d}\left(-F_{\left|\tilde{h}_{po}\right|^2}(y)\right)^d$$
$$= \sum_{d=0}^{j}\binom{j}{d}(-1)^d\left(F_{\left|\tilde{h}_{po}\right|^2}(y)\right)^d \tag{A.7}$$

2. Let us suppose, $X_1, X_2, ..., X_{M_p}$ are M_p independent variates, each with CDF $F_{\left|\tilde{h}_{po}\right|^2}(y)$, then CDF of kth order statistic in terms of unordered CDF function can be written as [216, Eq. (2.1.3)]

$$F_{|h_k^{po}|^2}(y) = \left(F_{\left|\tilde{h}_{po}\right|^2}(y)\right)^k \sum_{j=0}^{M_p-k}\binom{k+j-1}{k-1}\left(1-F_{\left|\tilde{h}_{po}\right|^2}(y)\right)^j, \tag{A.8}$$

where $|h|^2$ and $|\tilde{h}|^2$ denotes estimated ordered and estimated unordered channel gains, respectively. On substituting value from (A.7) in (A.8), we get

$$F_{|h_k^{po}|^2}(y) = \sum_{j=0}^{M_p-k}\binom{k+j-1}{k-1}\sum_{d=0}^{j}\binom{j}{d}(-1)^d\left(F_{\left|\tilde{h}_{po}\right|^2}(y)\right)^{d+k}$$
$$= \sum_{j=0}^{M_p-k}\sum_{d=0}^{j}\frac{(k+j-1)!}{j!\,(k-1)!}\times\frac{j!}{d!\,(j-d)!}(-1)^d\left(F_{\left|\tilde{h}_{po}\right|^2}(y)\right)^{d+k} \tag{A.9}$$

On exchanging the two summation, we can rewrite (A.9) as

$$F_{|h_k^{po}|^2}(y) = \frac{1}{(k-1)!}\sum_{d=0}^{M_p-k}\sum_{j=d}^{M_p-k}\frac{(k+j-1)!}{d!\,(j-d)!}\times(-1)^d\left(F_{\left|\tilde{h}_{po}\right|^2}(y)\right)^{d+k}$$
$$= \frac{1}{(k-1)!}\sum_{d=0}^{M_p-k}\frac{(-1)^d}{d!}\left(F_{\left|\tilde{h}_{po}\right|^2}(y)\right)^{d+k}\sum_{j=d}^{M_p-k}\frac{(k+j-1)!}{(j-d)!} \tag{A.10}$$

3. By the principle of mathematical induction, we know that

$$\sum_{j=d}^{M_p-k}\frac{(k+j-1)!}{(j-d)!} = \frac{M_p!}{(k+d)\left(M_p-k-d\right)!} \tag{A.11}$$

On substituting value from (A.11) in (A.10), we get

$$F_{|h_k^{po}|^2}(y) = \frac{1}{(k-1)!}\sum_{d=0}^{M_p-k}\frac{(-1)^d}{d!}\left(F_{|\tilde{h}_{po}|^2}(y)\right)^{d+k} \times \frac{M_p!}{(k+d)\left(M_p-k-d\right)!} \tag{A.12}$$

On multiplying $\left(M_p - k\right)!$ on numerator and denominator

$$\begin{aligned} F_{|h_k^{po}|^2}(y) &= \frac{1}{(k-1)!}\sum_{d=0}^{M_p-k}\frac{(-1)^d}{d!}\left(F_{|\tilde{h}_{po}|^2}(y)\right)^{d+k} \times \frac{M_p!}{(k+d)\left(M_p-k-d\right)!} \\ &\times \frac{\left(M_p-k\right)!}{\left(M_p-k\right)!} = \frac{M_p!}{(k-1)!\left(M_p-k\right)!}\sum_{d=0}^{M_p-k}\frac{(-1)^d}{(k+d)}\left(F_{|\tilde{h}_{po}|^2}(y)\right)^{d+k} \\ &\times \frac{\left(M_p-k\right)!}{d!\left(M_p-k-d\right)!} \end{aligned} \tag{A.13}$$

On substituting $\epsilon_k = \frac{M_p!}{(k-1)!(M_p-k)!}$, we get

$$F_{|h_k^{po}|^2}(y) = \epsilon_k \sum_{d=0}^{M_p-k}\binom{M_p-k}{d}\frac{(-1)^d}{k+d}\left(F_{|\tilde{h}_{po}|^2}(y)\right)^{k+d} \tag{A.14}$$

A.3 Proof of Proposition 2.3

The SINR of a typical MU, originally defined in (2.2), can be reformulated as $\gamma_m = \frac{H}{\sum_t \rho_t^{\mathcal{I}} \Upsilon_t + 1}$, where $H = \rho_m P_m |\tilde{h}_{m_o}|^2$. The ergodic rate for the MU is then given by $\mathcal{E}_m = \omega_m \times \mathbb{E}\left[\log_2(1+\gamma_m)\right]$. Using the identity from [217], the natural logarithm can be expressed as $\ln(1+\gamma_m) = \int_0^\infty \frac{1}{x}(1-e^{-\gamma_m x})e^{-x}dx$. Accordingly, the ergodic rate can be computed as

$$\begin{aligned} \mathcal{E}_m &= \omega_m \times \mathbb{E}\left[\log_2(1+\gamma_m)\right], \\ &\overset{(a)}{=} \frac{\omega_m}{\ln 2}\mathbb{E}\left[\int_0^\infty \frac{1}{x}\left(1-e^{-x\mathrm{H}}\right)e^{-x\left(\sum_t \rho_t^{\mathcal{I}}\Upsilon_t + 1\right)}dx\right], \\ &= \frac{\omega_m}{\ln 2}\int_0^\infty \frac{1}{x}\left(1-\mathbb{E}\left[e^{-x\mathrm{H}}\right]\right)\mathbb{E}\left[e^{-x\sum_t \rho_t^{\mathcal{I}}\Upsilon_t}\right]e^{-x}dx. \end{aligned} \tag{A.15}$$

Step (a) follows from the change of variable $x = \left(\sum_t \rho_t^{\mathcal{I}}\Upsilon_t + 1\right)x'$, which is subsequently substituted back by relabeling $x' \to x$. As a result, the ergodic rate for a typical MU is obtained as shown in (2.39). The LT of a random variable Z is defined as $\mathcal{L}_Z(s) = \int_0^\infty e^{-sz} f_Z(z)dz$, where $f_Z(z)$ represents the PDF of Z. To evaluate the LT of the unordered channel gain for the MBS tier, $\mathcal{L}_{|\tilde{h}_{mo}|^2}(s)$, we utilize the PDF derived from the CDF given in (2.22), is given as

$$f_{|\tilde{h}_{mo}|^2}(y) \simeq -\pi \lambda_m \mathcal{Y}_m^2 \sum_{\iota=0}^{I} b_\iota^m c_\iota^m e^{-c_\iota^m y}. \tag{A.16}$$

Therefore, the LT of the unordered channel gain for the MBS tier is derived as shown in (2.40), by applying the LT definition along with the PDF expression in (A.16). Similarly, the LT of the unordered channel gain for the PBS tier can be computed following the same approach. The PDF of the unordered channel gain for the PBS tier is obtained by differentiating its CDF, i.e., $f_{|\tilde{h}_{po}|^2}(y) = \frac{d}{dy}\left(F_{|\tilde{h}_{po}|^2}(y)\right)$, where the CDF is provided in (2.27), is given as

$$f_{|\tilde{h}_{po}|^2}(y) \simeq -\sum_{\iota=0}^{I} b_\iota^p c_\iota^p e^{-c_\iota^p y}. \tag{A.17}$$

Accordingly, the LT of the unordered channel gain for the PBS tier is obtained as given in (2.42), by applying the LT definition in conjunction with the PDF in (A.17). The ergodic rate for a typical PU operating without NOMA is then derived as shown in (2.41), utilizing the results from (2.42) and (2.24).

A.4 Proof of Proposition 2.4

The PDF of the ordered channel gain for the PBS tier can be obtained by differentiating its CDF, given in (2.34), as $f_{|h_k^{po}|^2}(y) = \frac{d}{dy}\left(F_{|h_k^{po}|^2}(y)\right)$, which is expressed as follows:

$$f_{|h_k^{po}|^2}(y) = -\epsilon_k \sum_{r=0}^{M_p-k} f_1 f_2 \sum_{T_k^r} f_3 \left(\prod_{\iota=0}^{I} (b_\iota^p)^{q_\iota}\right) \times \sum_{\iota=0}^{I} q_\iota c_\iota^p e^{-\sum_{\iota=0}^{I} q_\iota c_\iota^p y}. \tag{A.18}$$

The LT of order statistics $\left|h_k^{p_o}\right|^2$ can be expressed as provided in (2.44), using (A.18) and the LT formula. The SINR as specified in (2.6) can be rewritten as $\gamma_k^p = \frac{H_k}{H_l + \sum_\iota \rho_\iota^{\mathcal{I}} \Upsilon_\iota + 1}$, where $H_k = \rho_p P_p a_k \left|h_k^{p_o}\right|^2$, and $H_l = \rho_p P_p \left|h_k^{p_o}\right|^2 \sum_{l=k+1}^{M_p} a_l$. Following the same procedure as in (A.15) we get

$$
\begin{aligned}
\mathcal{E}_k^p &= \mathbb{E}\left[\log_2(1+\gamma_k^p)\right], \\
&= \frac{1}{\ln 2}\mathbb{E}\left[\int_0^\infty \frac{1}{x}\left(1-e^{-xH_k}\right)e^{-x\left(\sum_t \rho_t^{\mathcal{I}}\Upsilon_t + H_l + 1\right)}dx\right], \\
&= \frac{1}{\ln 2}\int_0^\infty \frac{1}{x}\left(1-\mathbb{E}\left[e^{-xH_k}\right]\right)\mathbb{E}\left[e^{-xH_l}\right]\mathbb{E}\left[e^{-xH_t}\right]e^{-x}\,dx, \\
&= \frac{1}{\ln 2}\int_0^\infty \frac{1}{x}\left(1-\mathbb{E}\left[e^{-xH_k}\right]\right)\prod_{l=k+1}^{M_p}\left(\mathbb{E}\left[e^{-x\hat{H}_l}\right]\right)\times\mathbb{E}\left[e^{-x\sum_t \rho_t^{\mathcal{I}}\Upsilon_t}\right]e^{-x}dx,
\end{aligned}
\tag{A.19}
$$

where $\hat{H}_l = \rho_p P_p a_l \left|h_k^{p_o}\right|^2$, while H_l and $H_t = \sum_t \rho_t^{\mathcal{I}}\Upsilon_t$ denote summation terms. Consequently, the ergodic rate for the kth user can be written as in (2.43).

A.5 Proof of Proposition 2.5

The SINR as specified in (2.18) can be rewritten as $\gamma_{k,ipSIC}^{hpa,p} = \frac{H_{k,ipSIC}^{hpa}}{H_{l,ipSIC}^{hpa}+\sum_t \rho_t^{\mathcal{I}}\Upsilon_t+1}$, where $H_{k,ipSIC}^{hpa} = \rho_p P_p a_k \left|h_k^{p_o}\right|^2 |\alpha_1|^2$, and $H_{l,ipSIC}^{hpa} = \rho_p P_p \left|h_k^{p_o}\right|^2 \Big[\sum_{\substack{l=1\\l\neq k}}^{M_p} a_l\beta_l\Big(|\alpha_1|^2 + |\alpha_3|^2 P_p^2\big|\sum_{i=1}^{M_p}\sqrt{a_i}\big|^4\Big) + |\alpha_3|^2 P_p^2 a_k\big|\sum_{i=1}^{M_p}\sqrt{a_i}\big|^4\Big]$.
Following the same procedure as in (A.19) we get,

$$
\begin{aligned}
\mathcal{E}_{k,ipSIC}^{hpa,p} &= \mathbb{E}\left[\log_2(1+\gamma_{k,ipSIC}^{hpa,p})\right], \\
&= \frac{1}{\ln 2}\int_0^\infty \frac{1}{x}\left(1-\mathbb{E}\left[e^{-xH_{k,ipSIC}^{hpa}}\right]\right)\prod_{\substack{l=1\\l\neq k}}^{M_p}\left(\mathbb{E}\left[e^{-x\hat{H}_{l,ipSIC}^{hpa}}\right]\right)\mathbb{E}\left[e^{-x\sum_t \rho_t^{\mathcal{I}}\Upsilon_t}\right]e^{-x}dx,
\end{aligned}
\tag{A.20}
$$

where $\hat{H}_{l,ipSIC}^{hpa} = \rho_p P_p \left|h_k^{p_o}\right|^2 \Big[a_l\beta_l\Big(|\alpha_1|^2 + |\alpha_3|^2 P_p^2\big|\sum_{i=1}^{M_p}\sqrt{a_i}\big|^4\Big) + |\alpha_3|^2 P_p^2 a_k\big|\sum_{i=1}^{M_p}\sqrt{a_i}\big|^4\Big]$. Thus, we can write the final ergodic rate at the kth user using NOMA with ipSIC under non-linear HPA distortion as expressed in (2.45).

A.6 Proof of Proposition 2.6

The data rate achieved by users under NOMA should be no less than that achieved under OMA, i.e., ($Rate^{NOMA} \geq Rate^{OMA}$).

$$\log_2(1 + \gamma_{j\to k}^{p,NOMA}) \geq \frac{1}{2}\log_2(1 + \gamma_p^{OMA}). \tag{A.21}$$

Substituting the values of γ_p^{OMA} and $\gamma_{j\to k}^{p,NOMA}$ from (2.3) and (2.5) and assuming $\frac{\rho_p P_p |h_k^{p_o}|^2}{\sum_\iota \rho_\iota^{\mathcal{I}} \Upsilon_\iota + 1} = \gamma$, we get

$$\log_2\left(1 + \frac{a_j\gamma}{\gamma \sum\limits_{l=j+1}^{M_p} a_l + 1}\right) \geq \log_2\sqrt{(1+\gamma)}. \tag{A.22}$$

By specializing to the two-user scenario and letting $j = 1$ correspond to the CEU, the above inequality reduces to a constraint on the power allocation coefficients.

$$a_{CEU}\gamma - \gamma\left\{\sqrt{(1+\gamma)} - 1\right\}a_{CCU} \geq \sqrt{(1+\gamma)} - 1. \tag{A.23}$$

Since the sum of all NOMA power coefficients equals one, substitute $a_{CCU} = 1 - a_{CEU}$ into the above expression and simplify to arrive at (2.47).

Appendix B
Proofs of Chapter 4

B.1 Proof of Proposition 4.1

Let the channel gain be denoted as $X = \hat{h}_1(1 + r_1^{\alpha})^{-1}$, $Y = \hat{h}_2(1 + r_2^{\alpha})^{-1}$, and $Z = \hat{h}_{12}(1 + r_{12}^{\alpha})^{-1}$. The cumulative distribution function (CDF) of channel gain Y can be expressed as

$$F_Y(y) = \frac{2}{\mathcal{Y}_f^2} \int_0^{\mathcal{Y}_f} \left(1 - e^{-\frac{\left(1+r_2^{\alpha_f}\right)y}{1+\left(1+r_2^{\alpha_f}\right)\sigma_e^2}} \right) r_2 dr_2. \tag{B.1}$$

Applying G-C quadrature [218], CDF can be approximated as

$$F_Y(y) \approx \frac{1}{2} \sum_{n=0}^{N} b_n e^{-\frac{c_n}{1+c_n\sigma_e^2} y}, \tag{B.2}$$

where $b_n = -\omega_N(1 + \theta_n)\sqrt{1 - \theta_n^2}$, $b_0 = -\sum_1^N b_n$, $\omega_N = \pi/N$, $\theta_n = \cos\left(\frac{2n-1}{2N}\pi\right)$, $c_n = 1 + \left(\frac{\mathcal{Y}_f}{2}\theta_n + \frac{\mathcal{Y}_f}{2}\right)^{\alpha}$, $c_0 = 0$, N is complexity-accuracy trade-off parameter for the G-C quadrature. Equation (4.10) can be rewritten in terms of random variable Y as

$$P_{cc}^1 = \mathcal{P}(Y < \epsilon_1) + \mathcal{P}(Y \geq \epsilon_1, Y < \epsilon_2). \tag{B.3}$$

Second part of (B.3) gives outage for condition $\epsilon_1 < \epsilon_2$. Thus, for the case when $\epsilon_1 \geq \epsilon_2$, the outage probability is given by the CDF of Y with $y = \epsilon_1(1 + \beta \sum_t \rho_t \mathcal{I}_t + \rho_f \sigma_e^2)$, and substituting y in (B.2) to get (4.11).

V. Bhatia et al., *Advanced NOMA Techniques for Heterogeneous Cellular Networks*, Wireless Networks, https://doi.org/10.1007/978-3-032-09551-0

B.2 Proof of Proposition 4.2

Let the first and the second term of (4.13) be denoted as Θ_1 and Θ_2. The terms of Θ_1 are independent. Therefore, Θ_1 can be obtained as

$$\Theta_1 = \mathcal{P}\left(\gamma_{2\to1}^1 < \phi_1\right)\mathcal{P}\left(\gamma_1^1 < \phi_1\right). \tag{B.4}$$

$\mathcal{P}\left(\gamma_{2\to1}^1 < \phi_1\right)$ is given by (4.11) and the analysis for $\mathcal{P}\left(\gamma_1^1 < \phi_1\right)$ is carried in the same manner, i.e., by assuming the channel gain $X = \hat{h}_1(1 + r_1^{\alpha})^{-1}$ and following the same steps as in the Appendix A. The expression of Θ_1, is given as

$$\Theta_1 = \frac{1}{4}\sum_{n=0}^{N}\sum_{k=0}^{K} b_n b_k e^{-\frac{(c_n+c_k)\epsilon_1}{(1+c_n\sigma_e^2)(1+c_k\sigma_e^2)}} \mathcal{L}_{I_p}[s_p]\mathcal{L}'_{I_p}[s_p] \tag{B.5}$$

where $\mathcal{L}_I[s]$ and $\mathcal{L}'_I[s]$ represents the Laplace transform of the interference at CEU during the direct phase transmission and cooperative phase transmission, respectively. Θ_2 can be written as

$$\Theta_2 = \underbrace{\mathcal{P}\left(\gamma_1^1 < \phi_1\right)}_{\Theta_{21}}\underbrace{\mathcal{P}\left(\gamma_{2\to1}^1 \geq \phi_1, \gamma_1^2 < \phi_1\right)}_{\Theta_{22}}. \tag{B.6}$$

The evaluation of Θ_{21} is completed in (B.5). Therefore, the expression of Θ_{21} is given as

$$\Theta_{21} = \frac{1}{2}\sum_{n=0}^{N} b_n e^{-\frac{c_n(\epsilon_1+\rho_f\sigma_e^2)}{(1+c_n\sigma_e^2)}} \mathcal{L}_{I_p}[s_p]. \tag{B.7}$$

Consider $\rho_f\sigma_e^2 + \rho_f\mathcal{I}_f + \rho_m\mathcal{I}_m = \mathcal{I}$, and $\epsilon_1' = \epsilon_1(\beta\mathcal{I} + 1)$, using this Θ_{22} can be given from (B.6) as

$$\Theta_{22} = \mathcal{P}\Big(Z < \frac{\phi_1(\mathcal{I}+1)}{\eta(1-\beta)(\rho y + \mathcal{I})}, Y \geq \epsilon_1'\Big). \tag{B.8}$$

Consider high SNR approximation, i.e., assuming $\rho \to \infty$, which approximate $\rho y + \mathcal{I} \approx \rho y$ and $\epsilon' \approx 0$. Using these approximations Θ_{22} can be rewritten as

$$\Theta_{22} = \int_0^{\mathcal{Y}_f}\int_0^{\mathcal{Y}_f}\underbrace{\int_0^{\infty} F_Z\left(\frac{\phi_1(\mathcal{I}+1)}{\eta(1-\beta)\rho y}\right) f_Y(y)dy}_{I_1}\, f_{\mathcal{Y}_f}(r_1)dr_1 f_{\mathcal{Y}_f}(r_2)dr_2. \tag{B.9}$$

Consider $Q = \frac{\phi_1}{\eta(1-\beta)\rho_f}$ and replacing the upper integral of I_1 to be B and take the limit as B$\to\infty$. Applying G-C Quadrature to find the approximation of I_1 as

$$I_1 \approx 1 - \lim_{B\to\infty} \frac{\pi B}{2N} \sum_{n=1}^{N} \frac{(1+r_2^{\alpha_f})}{(1+r_2^{\alpha_f}\sigma_e^2)} \left(e^{-\frac{2\left(1+r_{12}^{\alpha_f}\right)Q}{\left(1+r_{12}^{\alpha_f}\sigma_e^2\right)(1+\theta_n)B}} \right) \times$$

$$\mathcal{L}_{I_p}\left[s_p\right] e^{\frac{-\left(1+r_2^{\alpha_f}\right)(1+\theta_n)B}{2\left(1+r_2^{\alpha_f}\sigma_e^2\right)}} \left(1-\theta_n^2\right)^{\frac{1}{2}}, \tag{B.10}$$

where $s_p = \frac{2(1+r_{12}^{\alpha})Q}{(1+r_{12}^{\alpha}\sigma_e^2)(1+\theta_n)B}$, $\theta_n = cos\left(\frac{2n-1}{2N}\right)\pi$. Substituting the value of I_1 in (B.9), applying $f_{\mathcal{Y}_f}(r_1) = \frac{2r_1}{\mathcal{Y}_f^2}$, $f_{\mathcal{Y}_f}(r_2) = \frac{2r_2}{\mathcal{Y}_f^2}$ and approximating the distance as $r_{12} \approx r_1$, Θ_{22} is given as

$$\Theta_{22} = 1 - \lim_{B\to\infty} \frac{2\pi B}{N\mathcal{Y}_f^4} \sum_{n=1}^{N} \left(1-\theta_n^2\right)^{\frac{1}{2}} \underbrace{\int_0^{\mathcal{Y}_f} \frac{(1+r_2^{\alpha})}{(1+r_2^{\alpha}\sigma_e^2)} e^{-\frac{\left(1+r_2^{\alpha}\right)K_n}{\left(1+r_2^{\alpha}\sigma_e^2\right)}} r_2 dr_2}_{I_2} \times$$

$$\underbrace{\int_0^{\mathcal{Y}_f} \left(e^{-\frac{\left(1+r_1^{\alpha_f}\right)Q}{\left(1+r_1^{\alpha_f}\sigma_e^2\right)K_n}} \right) \mathcal{L}_{I_p}\left[s_p\right] r_1 dr_1}_{I_3} \tag{B.11}$$

where $K_n = \frac{(1+\theta_n)B}{2}$, $s_m = \frac{(1+r_1^{\alpha_f})Q_1\rho_m}{(1+r_1^{\alpha_f}\sigma_e^2)K_n}$, $s_f = \frac{(1+r_1^{\alpha_f})Q_1\rho_f}{(1+r_1^{\alpha_f}\sigma_e^2)K_n}$. After solving for I_2 and Â applying G-C quadrature to get approximation of I_3. The expression of Θ_{22} is given as

$$\Theta_{22} = 1 \quad \lim_{B\to\infty} \frac{\pi^2 B}{\alpha^2 NM\mathcal{Y}_f^3} \sum_{n=1}^{N} \left(1 - \theta_n^2\right)^{\frac{1}{2}} \frac{e^{-K_n}}{K_n^{1+\frac{2}{\alpha}}} \left((2+\alpha K_n)\Gamma\left(\frac{2}{\alpha}\right) - \alpha K_n \Gamma\left(\frac{2}{\alpha}, \mathcal{Y}_f^{\alpha} K_n\right) \right.$$

$$\left. \sum_{m=1}^{M} b_m e^{-\frac{c_m Q_1}{(1+c_m\sigma_e^2)K_n}} - \alpha\Gamma\left(\frac{2+\alpha}{\alpha}, \mathcal{Y}_f^{\alpha} K_n\right)\right) \mathcal{L}_{I_p}\left[\frac{d_m Q\rho_m}{(1+d_m\sigma_e^2)K_n}\right], \tag{B.12}$$

where Â $c_m = 1 + \left((1+\varphi_m)\frac{\mathcal{Y}_f}{2}\right)^{\alpha}$, $d_m = 1 + (1+\varphi_m)\frac{\mathcal{Y}_f}{2}$, $\varphi_m = \cos\left(\frac{2m-1}{2M}\pi\right)$ and M is the complex accuracy trade-off parameter. After substituting Θ_{21} and Θ_{22} from (B.7) and (B.12) into (B.6), respectively to get Θ_2. Then, substituting for Θ_1 and Θ_2 in (4.13) to get closed form expression for outage probability of CEU as given in (4.14).

Appendix C
Proofs of Chapter 5

C.1 Proof of Equation 5.10

Assuming a homogeneous PPP and utilizing polar coordinates, the CDF of the unordered channel gain in the PBS tier is formulated as [114]:

$$F_{|\ddot{h}_{p_o}|^2}(y) = \frac{2}{\mathcal{Y}_p^2} \int_0^{\mathcal{Y}_p} (1 - e^{-\left(1+r_p^{\nu_p}\right)y}) r_p dr_p. \tag{C.1}$$

Using the Gauss-Chebyshev quadrature method [146] on Eq. (C.1), we obtain

$$F_{|\ddot{h}_{p_o}|^2}(y) \simeq \sum_{q=0}^{Q} b_q^p e^{-c_q^p y}. \tag{C.2}$$

Thus, the OP, as expressed in (5.9), can be reformulated as,

$$\mathbb{P}_O^p \approx \sum_{q=0}^{Q} b_q^p e^{-c_q^p \left\{ \frac{\phi_p}{\rho_p P_p} \left(\sum_t \rho_t^t \mathbb{I}_t + \zeta_p \right) \right\}}. \tag{C.3}$$

By solving (C.3), the result leads directly to the expression given in (5.10).

C.2 Proof of Equation 5.14

As the successful self-message decoding in SIC directly impacts the OP, it can be expressed for the kth user as:

V. Bhatia et al., *Advanced NOMA Techniques for Heterogeneous Cellular Networks*, Wireless Networks, https://doi.org/10.1007/978-3-032-09551-0

$$\mathbb{P}_k^{O,p} = \Pr\left[|h_k^{p_o}|^2 < \frac{\psi_{max}\left(\sum_t \rho_t^t \mathbb{I}_t + \zeta_p\right)}{\rho_p P_p} \right]. \tag{C.4}$$

This results in the OP as $\mathbb{P}_k^{O,p} = F_{|h_k^{p_o}|^2}(y)$, where $y = \frac{\psi_{max}(\sum_t \rho_t^t \mathbb{I}_t + \zeta_p)}{\rho_p P_p}$ and $\psi_{max} = \max(\psi_1, \psi_2, \ldots, \psi_k)$. For a CEU when $k = 1$, OP is simply calculated with $y = \frac{\psi_j(\sum_t \rho_t^t \mathbb{I}_t + \zeta_p)}{\rho_p P_p}$, as it decodes its own message without performing SIC. $\psi_j = \frac{\phi_j}{a_j - \phi_j\left(\sum_{l=j+1}^{M_p} a_l\right)}$, where, ϕ_j is defined as $2^{R_j} - 1$, with R_j representing the target data rate for the jth user.

The connection between the CDF of ordered and unordered channel gains of the PBS tier is expressed as follows [114]:

$$F_{|h_k^{p_o}|^2}(y) = \epsilon_k \sum_{d=0}^{M_p-k} \binom{M_p - k}{d} \frac{(-1)^d}{k+d} \left(F_{|\ddot{h}_{p_o}|^2}(y)\right)^{k+d}. \tag{C.5}$$

Upon substituting (C.2) in (C.5) and employing the multinomial theorem, the CDF for the ordered channel gains is obtained as

$$F_{|h_k^{p_o}|^2}(y) \simeq \epsilon_k \sum_{d=0}^{M_p-k} \binom{M_p - k}{d} \frac{(-1)^d}{k+d} \sum_{T_k^d} \binom{k+d}{l_0 + \ldots + l_N} \prod_{q=0}^{Q} (b_q^p)^{l_q} e^{-\sum\limits_{q=0}^{Q} l_q c_q^p y}. \tag{C.6}$$

Upon substituting "y" into (C.6) and solving it yields the expression presented in (5.14).

References

1. A. Dogra, R.K. Jha, S. Jain, A survey on beyond 5G network with the advent of 6G: architecture and emerging technologies. IEEE Access 9:67 512–67 547 (2020)
2. P.K. Singya, N. Kumar, V. Bhatia, Mitigating NLD for wireless networks: effect of nonlinear power amplifiers on future wireless communication networks. IEEE Microw. Mag. **18**(5), 73–90 (2017)
3. 3rd Generation Partnership Project (3GPP), Technical specification group radio access network; evolved universal terrestrial radio access (E-UTRA); physical channels and modulation (Release 15) (2019)
4. J.M. Meredith, Study on downlink multiuser superposition transmission for LTE, in *TSG RAN Meeting*, vol. 67 (2015)
5. 3rd Generation Partnership Project (3GPP), Study on non-orthogonal multiple access (NOMA) for NR (Release 16) (2018)
6. K.S. Gilhousen, I.M. Jacobs, R. Padovani, A.J. Viterbi, L.A. Weaver, C.E. Wheatley, On the capacity of a cellular cdma system. IEEE Trans. Veh. Technol. **40**(2), 303–312 (1991)
7. J. Li, X. Wu, R. Laroia, *OFDMA Mobile Broadband Communications: A Systems Approach* (Cambridge University Press, 2013)
8. E. Access, Further advancements for E-UTRA physical layer aspects. 3GPP Tech. Specification TR **36**, V2 (2010)
9. Z. Ding, Y. Liu, J. Choi, Q. Sun, M. Elkashlan, I. Chih-Lin, H.V. Poor, Application of non-orthogonal multiple access in LTE and 5G networks. IEEE Commun. Mag. **55**(2), 185–191 (2017)
10. Y. Saito, A. Benjebbour, Y. Kishiyama, T. Nakamura, System-level performance evaluation of downlink non-orthogonal multiple access (NOMA), in *2013 IEEE 24th Annual International Symposium on Personal, Indoor, and Mobile Radio Communications (PIMRC)* (2013), pp. 611–615
11. R. Hoshyar, F.P. Wathan, R. Tafazolli, Novel low-density signature for synchronous CDMA systems over AWGN channel. IEEE Trans. Signal Process. **56**(4), 1616–1626 (2008)
12. Y. Du, B. Dong, Z. Chen, J. Fang, X. Wang, A fast convergence multiuser detection scheme for uplink SCMA systems. IEEE Wirel. Commun. Lett. **5**(4), 388–391 (2016)
13. Z. Yuan, G. Yu, W. Li, Y. Yuan, X. Wang, J. Xu, Multi-user shared access for internet of things, in *IEEE 83rd Vehicular Technology Conference (VTC Spring)* (IEEE, 2016), pp. 1–5
14. X. Dai, S. Chen, S. Sun, S. Kang, Y. Wang, Z. Shen, J. Xu, Successive interference cancelation amenable multiple access (SAMA) for future wireless communications, in *IEEE International Conference on Communication Systems* (IEEE, 2014), pp. 222–226

V. Bhatia et al., *Advanced NOMA Techniques for Heterogeneous Cellular Networks*, Wireless Networks, https://doi.org/10.1007/978-3-032-09551-0

15. M. Vaezi, G.A.A. Baduge, Y. Liu, A. Arafa, F. Fang, Z. Ding, Interplay between NOMA and other emerging technologies: a survey. IEEE Trans. Cognit. Commun. Netw. **5**(4), 900–919 (2019)
16. Y. Saito, Y. Kishiyama, A. Benjebbour, T. Nakamura, A. Li, K. Higuchi, Non-orthogonal multiple access (NOMA) for cellular future radio access, in *IEEE 77th Vehicular Technology Conference (VTC Spring)* (IEEE, 2013), pp. 1–5
17. Z. Ding, P. Fan, H.V. Poor, Impact of user pairing on 5G nonorthogonal multiple-access downlink transmissions. IEEE Trans. Veh. Technol. **65**(8), 6010–6023 (2015)
18. T. Cover, Broadcast channels. IEEE Trans. Inf. Theory **18**(1), 2–14 (1972)
19. T.M. Cover, *Elements of Information Theory* (Wiley, 1999)
20. P. Bergmans, Random coding theorem for broadcast channels with degraded components. IEEE Trans. Inf. Theory **19**(2), 197–207 (1973)
21. A. Carleial, Interference channels. IEEE Trans. Inf. Theory **24**(1), 60–70 (1978)
22. T. Cover, A.E. Gamal, Capacity theorems for the relay channel. IEEE Trans. Inf. Theory **25**(5), 572–584 (1979)
23. A.J. Grant, B. Rimoldi, R.L. Urbanke, P.A. Whiting, Rate-splitting multiple access for discrete memoryless channels. IEEE Trans. Inf. Theory **47**(3), 873–890 (2001)
24. I. Csiszár, J. Korner, Broadcast channels with confidential messages. IEEE Trans. Inf. Theory **24**(3), 339–348 (1978)
25. S. Vanka, S. Srinivasa, Z. Gong, P. Vizi, K. Stamatiou, M. Haenggi, Superposition coding strategies: design and experimental evaluation. IEEE Trans. Wirel. Commun. **11**(7), 2628–2639 (2012)
26. P. Bergmans, A simple converse for broadcast channels with additive white gaussian noise (corresp.). IEEE Trans. Inf. Theory **20**(2), 279–280 (1974)
27. J.G. Andrews, Interference cancellation for cellular systems: a contemporary overview. IEEE Wirel. Commun. **12**(2), 19–29 (2005)
28. P. Patel, J. Holtzman, Analysis of a simple successive interference cancellation scheme in a ds/cdma system. IEEE J. Sel. Areas Commun. **12**(5), 796–807 (1994)
29. P.W. Wolniansky, G.J. Foschini, G.D. Golden, R.A. Valenzuela, V-blast: An architecture for realizing very high data rates over the rich-scattering wireless channel, in *International Symposium on Signals, Systems, and Electronics* (IEEE, 1998), pp. 295–300
30. E. Gelal, J. Ning, K. Pelechrinis, T.-S. Kim, I. Broustis, S.V. Krishnamurthy, B.D. Rao, Topology control for effective interference cancellation in multiuser mimo networks. IEEE/ACM Trans. Netw. **21**(2), 455–468 (2012)
31. C. Jiang, Y. Shi, Y.T. Hou, W. Lou, S. Kompella, S.F. Midkiff, Squeezing the most out of interference: an optimization framework for joint interference exploitation and avoidance, in *IEEE INFOCOM* (IEEE, 2012), pp. 424–432
32. C. Xu, L. Ping, P. Wang, S. Chan, X. Lin, Decentralized power control for random access with successive interference cancellation. IEEE J. Sel. Areas Commun. **31**(11), 2387–2396 (2013)
33. J. Lee, J.G. Andrews, D. Hong, Spectrum-sharing transmission capacity with interference cancellation. IEEE Trans. Commun. **61**(1), 76–86 (2012)
34. A. Damnjanovic, J. Montojo, Y. Wei, T. Ji, T. Luo, M. Vajapeyam, T. Yoo, O. Song, D. Malladi, A survey on 3GPP heterogeneous networks. IEEE Wirel. Commun. **18**(3), 10–21 (2011)
35. H.S. Dhillon, R.K. Ganti, F. Baccelli, J.G. Andrews, Modeling and analysis of k-tier downlink heterogeneous cellular networks. IEEE J. Sel. Areas Commun. **30**(3), 550–560 (2012)
36. W. Bao, B. Liang, Stochastic analysis of uplink interference in two-tier femtocell networks: open versus closed access. IEEE Trans. Wirel. Commun. **14**(11), 6200–6215 (2015)
37. D.J. Daley, D. Vere-Jones et al., *An Introduction to the Theory of Point Processes: Volume I: Elementary Theory and Methods* (Springer, 2003)
38. A. Baddeley et al., Analysing spatial point patterns in r, Technical Report (2008)
39. R. Schneider, W. Weil, *Stochastic and Integral Geometry*, vol. 1 (Springer, 2008)
40. S.N. Chiu, D. Stoyan, W.S. Kendall, J. Mecke, *Stochastic Geometry and Its Applications* (Wiley, 2013)

41. M. Haenggi, J.G. Andrews, F. Baccelli, O. Dousse, M. Franceschetti, Stochastic geometry and random graphs for the analysis and design of wireless networks. IEEE J. Sel. Areas Commun. **27**(7), 1029–1046 (2009)
42. S. Weber, J.G. Andrews, N. Jindal, An overview of the transmission capacity of wireless networks. IEEE Trans. Commun. **58**(12), 3593–3604 (2010)
43. M. Haenggi, R.K. Ganti et al., Interference in large wireless networks. Found. Trends® Netw. 3(2):127–248 (2009)
44. F. Baccelli, B. Błaszczyszyn, Stochastic geometry and wireless networks, volume i–theory, volume 3, no 3–4 of foundations and trends in networking (2009)
45. F. Baccelli, S. Zuyev, Stochastic geometry models of mobile communication networks, in *Frontiers in Queueing: Models and Applications in Science and Engineering* (1998), pp. 227–243
46. F. Baccelli, M. Klein, M. Lebourges, S. Zuyev, Stochastic geometry and architecture of communication networks. Telecommun. Syst. **7**, 209–227 (1997)
47. J.G. Andrews, F. Baccelli, R.K. Ganti, A tractable approach to coverage and rate in cellular networks. IEEE Trans. Commun. **59**(11), 3122–3134 (2011)
48. S. Mukherjee, *Analytical Modeling of Heterogeneous Cellular Networks*. (Cambridge University Press, 2014)
49. S. Singh, H.S. Dhillon, J.G. Andrews, Offloading in heterogeneous networks: modeling, analysis, and design insights. IEEE Trans. Wirel. Commun. **12**(5), 2484–2497 (2013)
50. H.S. Dhillon, Y. Li, P. Nuggehalli, Z. Pi, J.G. Andrews, Fundamentals of heterogeneous cellular networks with energy harvesting. IEEE Trans. Wirel. Commun. **13**(5), 2782–2797 (2014)
51. Z. Gao, L. Dai, D. Mi, Z. Wang, M.A. Imran, M.Z. Shakir, MmWave massive-MIMO-based wireless backhaul for the 5G ultra-dense network. IEEE Wirel. Commun. **22**(5), 13–21 (2015)
52. J. Schloemann, H.S. Dhillon, R.M. Buehrer, A tractable analysis of the improvement in unique localizability through collaboration. IEEE Trans. Wirel. Commun. **15**(6), 3934–3948 (2016)
53. J.G. Andrews, T. Bai, M.N. Kulkarni, A. Alkhateeb, A.K. Gupta, R.W. Heath, Modeling and analyzing millimeter wave cellular systems. IEEE Trans. Commun. **65**(1), 403–430 (2016)
54. H. ElSawy, E. Hossain, M. Haenggi, Stochastic geometry for modeling, analysis, and design of multi-tier and cognitive cellular wireless networks: a survey. IEEE Commun. Surv. Tuts. **15**(3), 996–1019 (2013)
55. H. ElSawy, A. Sultan-Salem, M.-S. Alouini, M.Z. Win, Modeling and analysis of cellular networks using stochastic geometry: a tutorial. IEEE Commun. Surv. Tuts. **19**(1), 167–203 (2016)
56. Y. Liu, Z. Qin, M. Elkashlan, A. Nallanathan, J.A. McCann, Non-orthogonal multiple access in large-scale heterogeneous networks. IEEE J. Sel. Areas Commun. **35**(12), 2667–2680 (2017)
57. Z. Ding, L. Dai, H.V. Poor, MIMO-NOMA design for small packet transmission in the Internet of Things. IEEE Access **4**, 1393–1405 (2016)
58. H.-S. Jo, Y.J. Sang, P. Xia, J.G. Andrews, Heterogeneous cellular networks with flexible cell association: a comprehensive downlink SINR analysis. IEEE Trans. Wirel. Commun. **11**(10), 3484–3495 (2012)
59. J. Xu, J. Zhang, J.G. Andrews, On the accuracy of the wyner model in cellular networks. IEEE Trans. Wirel. Commun. **10**(9), 3098–3109 (2011)
60. P. Cardieri, Modeling interference in wireless ad hoc networks. IEEE Commun. Surv. Tuts. **12**(4), 551–572 (2010)
61. M. Haenggi, *Stochastic Geometry for Wireless Networks* (Cambridge University Press, 2012)
62. J.G. Andrews, F. Baccelli, R.K. Ganti, A tractable approach to coverage and rate in cellular networks. IEEE Trans. Commun. **59**(11), 3122–3134 (2011)
63. H. ElSawy, E. Hossain, A modified hard-core point process for analysis of random CSMA wireless networks in general fading environments. IEEE Trans. Commun. **61**(4), 1520–1534 (2013)
64. G. Alfano, M. Garetto, E. Leonardi, New insights into the stochastic geometry analysis of dense CSMA networks, in *Proceedings of IEEE INFOCOM* (2011), pp. 2642–2650

65. H.Q. Nguyen, F. Baccelli, D. Kofman, A stochastic geometry analysis of dense IEEE 802.11 networks, in *Proceedings of IEEE International Conference on Computing and Communications* (2007), pp. 1199–1207
66. M. Afshang, H.S. Dhillon, Fundamentals of modeling finite wireless networks using binomial point process. IEEE Trans. Wirel. Commun. **16**(5), 3355–3370 (2017)
67. M. Afshang, H.S. Dhillon, Poisson cluster process based analysis of HetNets with correlated user and base station locations. IEEE Trans. Wirel. Commun. **17**(4), 2417–2431 (2018)
68. C. Saha, H.S. Dhillon, N. Miyoshi, J.G. Andrews, Unified analysis of HetNets using poisson cluster processes under max-power association. IEEE Trans. Wirel. Commun. **18**(8), 3797–3812 (2019)
69. M.L. Ku, W. Li, Y. Chen, K.R. Liu, Advances in energy harvesting communications: past, present, and future challenges. IEEE Commun. Surv. Tuts. **18**(2), 1384–1412 (2015)
70. J. Huang, C.-C. Xing, C. Wang, Simultaneous wireless information and power transfer: technologies, applications, and research challenges. IEEE Commun. Mag. **55**(11), 26–32 (2017)
71. T.D.P. Perera, D.N.K. Jayakody, S.K. Sharma, S. Chatzinotas, J. Li, Simultaneous wireless information and power transfer (SWIPT): recent advances and future challenges. IEEE Commun. Surv. Tuts. **20**(1), 264–302 (2017)
72. A. Goldsmith, *Wireless Communications* (Cambridge University Press, 2005)
73. M.K. Simon, M.-S. Alouini, *Digital Communication over Fading Channels*, vol. 95 (Wiley, 2005)
74. Z. Ding, Z. Yang, P. Fan, H.V. Poor, On the performance of non-orthogonal multiple access in 5G systems with randomly deployed users. IEEE Signal Process. Lett. **21**(12), 1501–1505 (2014)
75. F. Fang, J. Cheng, Z. Ding, Joint energy efficient subchannel and power optimization for a downlink NOMA heterogeneous network. IEEE Trans. Veh. Technol. **68**(2), 1351–1364 (2018)
76. J. He, Z. Tang, Low-complexity user pairing and power allocation algorithm for 5G cellular network non-orthogonal multiple access. Electron. Lett. **53**(9), 626–627 (2017)
77. Z. Ding, M. Peng, H.V. Poor, Cooperative non-orthogonal multiple access in 5G systems. IEEE Commun. Lett. **19**(8), 1462–1465 (2015)
78. Y. Liu, Z. Ding, M. Elkashlan, J. Yuan, Non-orthogonal multiple access in large-scale underlay cognitive radio networks. IEEE Trans. Veh. Technol. **65**(12):10 152–10 157 (2016)
79. V.-D. Nguyen, H.D. Tuan, T.Q. Duong, O.-S. Shin, H.V. Poor, Joint fractional time allocation and beamforming for downlink multiuser MISO systems. IEEE Commun. Lett. **21**(12), 2650–2653 (2017)
80. J.A. Oviedo, H.R. Sadjadpour, A fair power allocation approach to NOMA in multiuser SISO systems. IEEE Trans. Veh. Technol. **66**(9), 7974–7985 (2017)
81. J. Choi, NOMA-based random access with multichannel ALOHA. IEEE J. Sel. Areas Commun. **35**(12), 2736–2743 (2017)
82. Y.-G. Lim, C.-B. Chae, G. Caire, Performance analysis of massive MIMO for cell-boundary users. IEEE Trans. Wirel. Commun. **14**(12), 6827–6842 (2015)
83. Y. Liu, M. Elkashlan, Z. Ding, G.K. Karagiannidis, Fairness of user clustering in MIMO non-orthogonal multiple access systems. IEEE Commun. Lett. **20**(7), 1465–1468 (2016)
84. Y. Liu, Z. Ding, M. Elkashlan, H.V. Poor, Cooperative non-orthogonal multiple access in 5G systems with SWIPT, in *23rd European Signal Processing Conference (EUSIPCO)* (IEEE, 2015), pp. 1999–2003
85. S. Timotheou, I. Krikidis, Fairness for non-orthogonal multiple access in 5G systems. IEEE Signal Process. Lett. **22**(10), 1647–1651 (2015)
86. Z. Yang, Z. Ding, P. Fan, G.K. Karagiannidis, On the performance of non-orthogonal multiple access systems with partial channel information. IEEE Trans. Commun. **64**(2), 654–667 (2015)
87. Z. Zhang, H. Sun, R.Q. Hu, Y. Qian, Stochastic geometry based performance study on 5g non-orthogonal multiple access scheme, in *IEEE Global Communications Conference (GLOBECOM)* (IEEE, 2016), pp. 1–6

88. J. Men, J. Ge, Performance analysis of non-orthogonal multiple access in downlink cooperative network. IET Commun. **9**(18), 2267–2273 (2015)
89. J. Men, J. Ge, Non-orthogonal multiple access for multiple-antenna relaying networks. IEEE Commun. Lett. **19**(10), 1686–1689 (2015)
90. Z. Ding, H. Dai, H.V. Poor, Relay selection for cooperative noma. IEEE Wirel. Commun. Lett. **5**(4), 416–419 (2016)
91. Y. Tian, A.R. Nix, M. Beach, On the performance of opportunistic NOMA in downlink CoMP networks. IEEE Commun. Lett. **20**(5), 998–1001 (2016)
92. J.-B. Kim, I.-H. Lee, Non-orthogonal multiple access in coordinated direct and relay transmission. IEEE Commun. Lett. **19**(11), 2037–2040 (2015)
93. Q.-T. Vien, N. Ogbonna, H.X. Nguyen, R. Trestian, P. Shah, Non-orthogonal multiple access for wireless downlink in cloud radio access networks, in *Proceedings of European Wireless Conference* (VDE, 2015), pp. 1–6
94. N. Shinohara, Power without wires. IEEE Microw. Mag. **12**(7), S64–S73 (2011)
95. P. Coporation, RF energy harvesting and wireless power for low-power applications. Pittsburgh (2011)
96. R. Vullers, R. van Schaijk, I. Doms, C. Van Hoof, R. Mertens, Micropower energy harvesting. Solid State Electron. **53**(7), 684–693 (2009)
97. P. Cheng, S. He, F. Jiang, Y. Gu, J. Chen, Optimal scheduling for quality of monitoring in wireless rechargeable sensor networks. IEEE Trans. Wirel. Commun. **12**(6), 3072–3084 (2013)
98. S. He, J. Chen, F. Jiang, D.K. Yau, G. Xing, Y. Sun, Energy provisioning in wireless rechargeable sensor networks. IEEE Trans. Mobile Comput. **12**(10), 1931–1942 (2012)
99. L.R. Varshney, Transporting information and energy simultaneously, in *Proceedings of IEEE International Symposium on Information Theory (ISIT)* (Toronto, ON, Canada, 2008), pp. 1612–1616
100. R. Zhang, C.K. Ho, MIMO broadcasting for simultaneous wireless information and power transfer. IEEE Trans. Wirel. Commun. **12**(5), 1989–2001 (2013)
101. A.A. Nasir, X. Zhou, S. Durrani, R.A. Kennedy, Relaying protocols for wireless energy harvesting and information processing. IEEE Trans. Wirel. Commun. **12**(7), 3622–3636 (2013)
102. I. Krikidis, S. Sasaki, S. Timotheou, Z. Ding, A low complexity antenna switching for joint wireless information and energy transfer in MIMO relay channels. IEEE Trans. Wirel. Commun. **62**(5), 1577–1587 (2014)
103. Y. Liu, S.A. Mousavifar, Y. Deng, C. Leung, M. Elkashlan, Wireless energy harvesting in a cognitive relay network. IEEE Trans. Wirel. Commun. **15**(4), 2498–2508 (2015)
104. Z. Ding, I. Krikidis, B. Sharif, H.V. Poor, Wireless information and power transfer in cooperative networks with spatially random relays. IEEE Trans. Wirel. Commun. **13**(8), 4440–4453 (2014)
105. P.D. Diamantoulakis, K.N. Pappi, Z. Ding, G.K. Karagiannidis, Wireless-powered communications with non-orthogonal multiple access. IEEE Trans. Wirel. Commun. **15**(12), 8422–8436 (2016)
106. Z. Ding, X. Lei, G.K. Karagiannidis, R. Schober, J. Yuan, V.K. Bhargava, A survey on non-orthogonal multiple access for 5G networks: research challenges and future trends. IEEE J. Sel. Areas Commun. **35**(10), 2181–2195 (2017)
107. Y. Liu, S. Zhang, X. Mu, Z. Ding, R. Schober, N. Al-Dhahir, E. Hossain, X. Shen, Evolution of NOMA toward next generation multiple access (NGMA) for 6G. IEEE J. Sel. Areas Commun. **40**, 1037–1071 (2022)
108. V. Bhatia, P. Swami, S. Sharma, R. Mitra, Non-orthogonal multiple access: an enabler for massive connectivity. J. Indian Inst. Sci. **100**(2), 337–348 (2020)
109. C. Zhao, Y. Liu, Y. Cai, M. Zhao, Z. Ding, Non-orthogonal multiple access for UAV-aided heterogeneous networks: a stochastic geometry model. IEEE Trans. Veh. Technol. **72**(1), 940–956 (2023)
110. Z. Ding, Z. Yang, P. Fan, H.V. Poor, On the performance of non-orthogonal multiple access in 5G systems with randomly deployed users. IEEE Signal Process. Lett. **21**(12), 1501–1505 (2014)

111. B. Makki, K. Chitti, A. Behravan, M.-S. Alouini, A survey of NOMA: current status and open research challenges. IEEE Open J. Commun. Soc. **1**, 179–189 (2020)
112. A.S. Parihar, P. Swami, V. Bhatia, Z. Ding, Performance analysis of SWIPT enabled cooperative-NOMA in heterogeneous networks using carrier sensing. IEEE Trans. Veh. Technol. **70**(10), 10 646–10 656 (2021)
113. S. Dong, J. Zhan, W. Hu, A. Mohajer, M. Bavaghar, A. Mirzaei, *Energy-efficient hierarchical resource allocation in uplink-downlink decoupled NOMA HetNets* (IEEE Trans. Netw. Serv, Manag, 2023)
114. P. Swami, V. Bhatia, S. Vuppala, T. Ratnarajah, A cooperation scheme for user fairness and performance enhancement in NOMA-HCN. IEEE Trans. Veh. Technol. **67**(12), 11 965–11 978 (2018)
115. M.K. Mishra, A. Trivedi, Spectral efficiency and deployment cost efficiency analysis of mmW/UHF-based cellular network. IEEE Trans. Veh. Technol. **68**(7), 6565–6577 (2019)
116. S. Ghosh, T. Acharya, S.P. Maity, On outage analysis in SWIPT enabled bidirectional D2D communications using spectrum sharing in cellular networks. IEEE Trans. Veh. Technol. **69**(9), 10 167–10 176 (2020)
117. I.A. Hemadeh, K. Satyanarayana, M. El-Hajjar, L. Hanzo, Millimeter-wave communications: physical channel models, design considerations, antenna constructions, and link-budget. IEEE Commun. Surv. Tuts. **20**(2), 870–913, 2nd Quart (2018)
118. C. Wu, X. Cai, J. Sheng, Z. Tang, B. Ai, Y. Wang, Parameter adaptation and situation awareness of LTE-R handover for high-speed railway communication. IEEE Trans. Intell. Transp. Syst. **23**(3), 1767–1781 (2022)
119. T. Qi, S. He, Power Up potential power amplifier technologies for 5G applications. IEEE Microw. Mag. **20**(6), 89–101 (2019)
120. E. Balti, M. Guizani, Impact of non-linear high-power amplifiers on cooperative relaying systems. IEEE Trans. Commun. **65**(10), 4163–4175 (2017)
121. D. Kumar, P.K. Singya, V. Bhatia, Impact of NLPA on SWIPT enabled two-way AF cooperative network, in *Proceedings of IEEE Vehicular Technology Conference (VTC)* (2021), pp. 1–5
122. M. Wildemeersch, T.Q. Quek, M. Kountouris, A. Rabbachin, C.H. Slump, Successive interference cancellation in heterogeneous networks. IEEE Trans. Commun. **62**(12), 4440–4453 (2014)
123. Z. Ding, R. Schober, H.V. Poor, Unveiling the importance of SIC in NOMA systems-Part 1: State of the art and recent findings. IEEE Commun. Lett. **24**(11), 2373–2377 (2020)
124. J. Andrews, T. Meng, Optimum power control for successive interference cancellation with imperfect channel estimation. IEEE Trans. Wireless Commun. **2**(2), 375–383 (2003)
125. G. Im, J.H. Lee, Outage probability for cooperative NOMA systems with imperfect SIC in cognitive radio networks. IEEE Commun. Lett. **23**(4), 692–695 (2019)
126. A. Celik, M.-C. Tsai, R.M. Radaydeh, F.S. Al-Qahtani, M.-S. Alouini, Distributed user clustering and resource allocation for imperfect NOMA in heterogeneous networks. IEEE Trans. Commun. **67**(10), 7211–7227 (2019)
127. S. Bisen, P. Shaik, V. Bhatia, On performance of energy harvested cooperative NOMA under imperfect CSI and imperfect SIC. IEEE Trans. Veh. Technol. **70**(9), 8993–9005 (2021)
128. A. Jee, S. Prakriya, Performance of energy and spectrally efficient AF relay-aided incremental CDRT NOMA based IoT network with imperfect SIC for smart cities. IEEE Internet Things J. (2022)
129. S. Ghosh, A. Al-Dweik, M.-S. Alouini, On the performance of end-to-end cooperative NOMA-based IoT networks with wireless energy harvesting. IEEE Internet Things J. **10**(18), 16 253–16 270 (2023)
130. P. Swami, M.K. Mishra, V. Bhatia, T. Ratnarajah, A. Trivedi, Performance analysis of sub-6 GHz/mmWave NOMA hybrid-HetNets using partial CSI. IEEE Trans. Veh. Technol. **71**,(12), 12 958–12 971 (2022)
131. S. Singh, M. Bansal, Outage analysis of NOMA-based cooperative relay systems with imperfect SIC. Phys. Commun. **43**(101219) (2020)

132. J. Venkataraman, M. Haenggi, O. Collins, Shot noise models for outage and throughput analyses in wireless ad hoc networks, in *Proceedings of IEEE Military Communications Conference*(2006), pp. 1–7
133. Y. Hmamouche, M. Benjillali, S. Saoudi, H. Yanikomeroglu, M. Di Renzo, New trends in stochastic geometry for wireless networks: a tutorial and survey. Proc. IEEE **109**(7), 1200–1252 (2021)
134. D. Stoyan, W.S. Kendall, S.N. Chiu, J. Mecke, *Stochastic Geometry and Its Applications* (Wiley, 2013)
135. M. Vaezi, R. Schober, Z. Ding, H.V. Poor, Non-orthogonal multiple access: common myths and critical questions. IEEE Wirel. Commun. **26**(5), 174–180 (2019)
136. S. Li, M. Derakhshani, S. Lambotharan, Outage-constrained robust power allocation for downlink MC-NOMA with imperfect SIC, in *018 IEEE International Conference on Communications (ICC)* (IEEE ICC, Kansas City, MO, USA, 2018), pp. 1–7
137. L. Ding, G.T. Zhou, Effects of even-order nonlinear terms on power amplifier modeling and predistortion linearization. IEEE Trans. Veh. Technol. **53**(1), 156–162 (2004)
138. O.B.H. Belkacem, M.L. Ammari, R. Dinis, Performance analysis of NOMA in 5G systems with HPA nonlinearities. IEEE Access **8**, 158 327–158 334 (2020)
139. J. Stoer, R. Bulirsch, *Introduction to Numerical Analysis* (Springer Science, Business Media, 2013)
140. Y. Liu, Z. Ding, M. Elkashlan, H.V. Poor, Cooperative non-orthogonal multiple access with simultaneous wireless information and power transfer. IEEE J. Sel. Areas Commun. **34**(4), 938–953 (2016)
141. Y. Liu, Z. Ding, M. Elkashlan, J. Yuan, Nonorthogonal multiple access in large-scale underlay cognitive radio networks. IEEE Trans. Veh. Technol. **65**(12), 10 152–10 157 (2016)
142. V. Kumar, B. Cardiff, S. Prakriya, M.F. Flanagan, Delay violation probability and effective rate of downlink NOMA over α-μ fading channels. IEEE Trans. Veh. Technol. **69**(10), 11 241–11 252 (2020)
143. E. Björnson, E.G. Larsson, How energy-efficient can a wireless communication system become? in *Proceedings of 52nd Asilomar Conference on Signals, Systems, and Computers* (2018), pp. 1252–1256
144. X. Yue, Y. Liu, S. Kang, A. Nallanathan, Z. Ding, Exploiting full/half-duplex user relaying in NOMA systems. IEEE Trans. Commun. **66**(2), 560–575 (2018)
145. A.S. Parihar, P. Swami, K. Choi, P. Brida, V. Bhatia, On performance of NOMA-based wireless powered communication networks assisted with power beacons and PPP distributed. IEEE Wirel. Commun. Lett. **12**(7), 1294–1298 (2023)
146. A. Baghel, A.S. Parihar, V. Bhatia, K. Choi, A.K. Mishra, On joint impact of HPA nonlinearity and imperfect SIC in NOMA enabled HCN using stochastic geometry. IEEE Trans. Veh. Technol. **73**(6), 8055–8068 (2024)
147. M. Katwe, K. Singh, C.-P. Li, Z. Ding, Spectral-efficient downlink systems under imperfect SIC and CSI: MC-NOMA or partial NOMA? IEEE Wirel. Commun. Lett. **13**(1), 133–137 (2024)
148. S. Singh, M. Bansal, On the outage performance of overlay cognitive STBC-NOMA system with imperfect SIC. IEEE Wirel. Commun. Lett. **10**(11), 2587–2591 (2021)
149. S. Soni, R. Makkar, D. Rawal, N. Sharma, Performance of selective DF-based multiple relayed NOMA system with imperfect CSI and SIC errors. IEEE Trans. Green Commun. Netw. **8**(1), 79–89 (2024)
150. Z. Ding, Y. Liu, J. Choi, Q. Sun, M. Elkashlan, I. Chih-Lin, H.V. Poor, Application of non-orthogonal multiple access in LTE and 5G networks. IEEE Commun. Mag. **55**(2), 185–191 (2017)
151. L. Dai, B. Wang, Y. Yuan, S. Han, I. Chih-Lin, Z. Wang, Non-orthogonal multiple access for 5G: solutions, challenges, opportunities, and future research trends. IEEE Commun. Mag. **53**(9), 74–81 (2015)
152. P. Grover, A. Sahai, Shannon meets tesla: wireless information and power transfer, in *Proceedings of IEEE International Symposium on Information Theory (ISIT)* (2010), pp. 2363–2367

153. D.W.K. Ng, E.S. Lo, R. Schober, Wireless information and power transfer: energy efficiency optimization in OFDMA systems. IEEE Trans. Wirel. Commun. **12**(12), 6352–6370 (2013)
154. B. Xu, Y. Chen, J.R. Carrión, T. Zhang, Resource allocation in energy-cooperation enabled two-tier NOMA HetNets toward green 5G. IEEE J. Sel. Areas Commun. **35**(12), 2758–2770 (2017)
155. Y. Cao, N. Zhao, G. Pan, Y. Chen, L. Fan, M. Jin, M.-S. Alouini, Secrecy analysis for cooperative NOMA networks with multi-antenna full-duplex relay. IEEE Trans. Commun. **67**(8), 5574–5587 (2019)
156. B. Chen, Y. Chen, Y. Cao, Y. Chen, N. Zhao, Z. Ding, X. Wang, Security enhancement using a novel two-slot cooperative NOMA scheme. IEEE Trans. Veh. Technol. **69**(3), 3470–3475 (2020)
157. J. Tang, J. Luo, M. Liu, D.K. So, E. Alsusa, G. Chen, K.-K. Wong, J.A. Chambers, Energy efficiency optimization for NOMA with SWIPT. IEEE J. Sel. Topics Signal Process. **13**(3), 452–466 (2019)
158. A. Rezaei, P. Azmi, N.M. Yamchi, M.R. Javan, H. Yanikomeroglu, Robust resource allocation for cooperative MISO-NOMA-based heterogeneous networks. IEEE Trans. Commun. **69**(6), 3864–3878 (2021)
159. Y. Liu, Z. Ding, M. Elkashlan, H.V. Poor, Cooperative non-orthogonal multiple access with simultaneous wireless information and power transfer. IEEE J. Sel. Areas Commun. **34**(4), 938–953 (2016)
160. Y. Liu, Z. Ding, M. Elkashlan, J. Yuan, Non-orthogonal multiple access in large-scale underlay cognitive radio networks. IEEE Trans. Veh. Technol. **65**(12), 10 152–10 157 (2016)
161. Z. Zhang, H. Sun, R.Q. Hu, Downlink and uplink non-orthogonal multiple access in a dense wireless network. IEEE J. Sel. Areas Commun. **35**(12), 2771–2784 (2017)
162. J. Wang, B. Xia, K. Xiao, Z. Chen, Performance analysis and power allocation strategy for downlink NOMA systems in large-scale cellular networks. IEEE Trans. Veh. Technol. **69**(3), 3459–3464 (2020)
163. S. Bisen, P. Shaik, V. Bhatia, On performance of energy harvested cooperative noma under imperfect csi and imperfect sic. IEEE Trans. Veh. Technol. **70**(9), 8993–9005 (2021)
164. P. Swami, V. Bhatia, S. Vuppala, T. Ratnarajah, Performance analysis of offloading in noma-hetnets using imperfect csi, in *IEEE ANTS* (IEEE, 2018), pp. 1–6
165. P. Shaik, P.K. Singya, N. Kumar, K.K. Garg, V. Bhatia, On impact of imperfect CSI over SWIPT device-to-device (D2D) MIMO relay systems, in *Proceedings of IEEE International Conference on Signal Processing and Communication Systems (SPCOM)* (2020), pp. 1–5
166. D. Kumar, P.K. Singya, V. Bhatia, *ASER analysis of hybrid receiver based SWIPT two-way relay network* (IEEE Trans. Veh, Technol, 2021)
167. I. Krikidis, Simultaneous information and energy transfer in large-scale networks with/without relaying. IEEE Trans. Commun. **62**(3), 900–912 (2014)
168. Z. Chen, J. Yuan, B. Vucetic, Analysis of transmit antenna selection/maximal-ratio combining in Rayleigh fading channels. IEEE Trans. Veh. Technol. **54**(4), 1312–1321 (2005)
169. N. Kong, Performance comparison among conventional selection combining, optimum selection combining and maximal ratio combining, in *Proceedings of IEEE International Conference on Communication* (2009), pp. 1–6
170. Z. Ding, P. Fan, H.V. Poor, Impact of user pairing on 5G nonorthogonal multiple-access downlink transmissions. IEEE Trans. Veh. Technol. **65**(8), 6010–6023 (2016)
171. F. Kara, H. Kaya, Improved user fairness in decode-forward relaying non-orthogonal multiple access schemes with imperfect SIC and CSI. IEEE Access **8**, 97 540–97 556 (2020)
172. A.S. Parihar, P. Swami, V. Bhatia, On performance of SWIPT enabled PPP distributed cooperative NOMA networks using stochastic geometry. IEEE Trans. Veh. Technol. **71**(5), 5639–5644 (2022)
173. A.S. Parihar, K. Singh, V. Bhatia, C.-P. Li, T.Q. Duong, Performance analysis of NOMA enabled active RIS-aided MIMO heterogeneous IoT networks with integrated sensing and communication. IEEE Internet Things J. 1 (2024)

174. M. Cheffena, Industrial wireless sensor networks: channel modeling and performance evaluation. EURASIP J. Wirel. Commun. Netw. **2012**(1), 1–8 (2012)
175. T. Bai, H. Zhang, J. Wang, C. Xu, M. Elkashlan, A. Nallanathan, L. Hanzo, Fifty years of noise modeling and mitigation in power-line communications. IEEE Commun. Surv. Tuts. **23**(1), 41–69 (2021)
176. K. Blackard, T. Rappaport, C. Bostian, Measurements and models of radio frequency impulsive noise for indoor wireless communications. IEEE J. Sel. Areas Commun. **11**(7), 991–1001 (1993)
177. B. Selim, M.S. Alam, J.V.C. Evangelista, G. Kaddoum, B.L. Agba, NOMA-based IoT networks: impulsive noise effects and mitigation. IEEE Commun. Mag. **58**(11), 69–75 (2020)
178. B. Selim, M.S. Alam, G. Kaddoum, B.L. Agba, Effect of impulsive noise on uplink NOMA systems. IEEE Trans. Veh. Technol. **69**(3), 3454–3458 (2020)
179. M. Hussain, H. Shakir, H. Rasheed, Deep learning approaches for impulse noise mitigation and classification in NOMA-based systems. IEEE Access **9**, 143 836–143 846 (2021)
180. A. Samir, M. Elsayed, A.A.A. El-Banna, K. Wu, B.M. ElHalawany, Performance of NOMA-based dual-hop hybrid powerline-wireless communication systems. IEEE Trans. Veh. Technol. **71**(6), 6548–6558 (2022)
181. M.S. Alam, B. Selim, I. Ahmed, G. Kaddoum, H. Yanikomeroglu, Bursty impulsive noise mitigation in NOMA: a map receiver-based approach. IEEE Commun. Lett. **25**(9), 2790–2794 (2021)
182. E.C. Strinati, G.C. Alexandropoulos, H. Wymeersch, B. Denis, V. Sciancalepore, R. D'Errico, A. Clemente, D.-T. Phan-Huy, E. De Carvalho, P. Popovski, Reconfigurable, intelligent, and sustainable wireless environments for 6G smart connectivity. IEEE Commun. Mag. **59**(10), 99–105 (2021)
183. D. Kumar, C.K. Singh, O.L.A. López, V. Bhatia, M. Latva-aho, Performance analysis of passive/active RIS aided wireless-powered IoT network with nonlinear energy harvesting. IEEE Trans. Wirel. Commun. (2024). https://doi.org/10.1109/TWC.2024.3505296
184. Z. Zhang, L. Dai, X. Chen, C. Liu, F. Yang, R. Schober, H.V. Poor, Active RIS versus passive RIS: which will prevail in 6G? IEEE Trans. Commun. **71**(3), 1707–1725 (2023)
185. K.K. Kishor, S.V. Hum, An amplifying reconfigurable reflectarray antenna. IEEE Trans. Antennas Propag. **60**(1), 197–205 (2012)
186. Z. Ding, H.V. Poor, A simple design of IRS-NOMA transmission. IEEE Commun. Lett. **24**(5), 1119–1123 (2020)
187. Z. Ding, R. Schober, H.V. Poor, On the impact of phase shifting designs on IRS-NOMA. IEEE Wirel. Commun. Lett. **9**(10), 1596–1600 (2020)
188. Y. Cheng, K.H. Li, Y. Liu, K.C. Teh, H.V. Poor, Downlink and uplink intelligent reflecting surface aided networks: NOMA and OMA. IEEE Trans. Wirel. Commun. **20**(6), 3988–4000 (2021)
189. M. Elhattab, M.A. Arfaoui, C. Assi, A. Ghrayeb, RIS-assisted joint transmission in a two-cell downlink NOMA cellular system. IEEE J. Sel. Areas Commun. **40**(4), 1270–1286 (2022)
190. M.H.N. Shaikh, V.A. Bohara, A. Srivastava, G. Ghatak, A downlink RIS-aided NOMA system with hardware impairments: performance characterization and analysis. IEEE Open J. Signal Process. **3**, 288–305 (2022)
191. X. Gu, G. Zhang, B. Zhuo, W. Duan, J. Wang, M. Wen, P.-H. Ho, On the performance of cooperative NOMA downlink: a RIS-aided D2D perspective. IEEE Trans. Cog. Commun. Netw. **9**(6), 1610–1624 (2023)
192. D. Kumar, P.K. Singya, O. Krejcar, K. Choi, V. Bhatia, Performance of IRS-aided FD two-way communication network with imperfect SIC. IEEE Tran. Veh. Technol. **72**(4), 5491–5496 (2023)
193. D. Kumar, C.K. Singh, O.L.A. López, V. Bhatia, M. Latva-aho, Performance analysis of active RIS-assisted downlink NOMA with transmit antenna selection. IEEE Veh. Technol. (2024). https://doi.org/10.1109/TVT.2025.3528379
194. Z. Shi, H. Lu, X. Xie, H. Yang, C. Huang, J. Cai, Z. Ding, Active RIS-aided EH-NOMA networks: a deep reinforcement learning approach. IEEE Trans. Commun. **71**(10), 5846–5861 (2023)

195. C. Gong, H. Li, S. Hao, K. Long, X. Dai, Active RIS enabled secure NOMA communications with discrete phase shifting. IEEE Trans. Wireless Commun. 1 (2023)
196. X. Yue et al., Exploiting active RIS in NOMA networks with hardware impairments. IEEE Trans. Veh. Technol. **73**(6), 8207–8221 (2024)
197. I.S. Gradshteyn, I.M. Ryzhik, *Table of Integrals, Series, and Products*, 8th edn. (Academic, New York, NY, USA, 2014)
198. D. Kumar, P.K. Singya, V. Bhatia, Performance analysis of hybrid two-way relay network with NLPA and hardware impairments, in *IEEE Wireless Commun. Networ. Confer. (WCNC)* (IEEE, 2021), pp. 1–6
199. S. Bisen, P. Shaik, V. Bhatia, On performance of energy harvested cooperative NOMA under imperfect CSI and imperfect SIC. IEEE Trans. Veh. Technol. **70**(9), 8993–9005 (2021)
200. D. Kumar, P.K. Singya, V. Bhatia, ASER analysis of hybrid receiver based SWIPT two-way relay network. *IEEE Trans. Veh. Technol.* **70**(10), 10 018–10 030 (2021)
201. D. Kumar, P.K. Singya, O. Krejcar, K. Choi, V. Bhatia, On performance of intelligent reflecting surface aided wireless powered IoT network with HIs. IEEE Commun. Lett. **27**(2), 502–506 (2023)
202. *The Wolfram Function Site*. http://functions.wolfram.com/
203. P.K. Singya, M.-S. Alouini, Performance of UAV-assisted multiuser terrestrial-satellite communication system over mixed FSO/RF channels. IEEE Trans. Aer. Electr. Sys. **58**(2), 781–796 (2021)
204. N. Kumar, P.K. Singya, V. Bhatia, ASER analysis of hexagonal and rectangular QAM schemes in multiple-relay networks. IEEE Trans. Veh. Technol. **67**(2), 1815–1819 (2018)
205. P.K. Singya, N. Kumar, V. Bhatia, Impact of imperfect CSI on ASER of hexagonal and rectangular QAM for AF relaying network. IEEE Commun. Lett. **22**(2), 428–431 (2018)
206. P.K. Singya, N. Kumar, V. Bhatia, M.-S. Alouini, On the performance analysis of higher order QAM schemes over mixed RF/FSO systems. IEEE Trans. Veh. Technol. **69**(7), 7366–7378 (2020)
207. M. Abramowitz, I.A. Stegun, *Handbook of Mathematical Functions with Formulas, Graphs, and Mathematical Tables*, 9th edn. (Dover, New York, NY, USA, 1970)
208. D. Sadhwani, R.N. Yadav, A simplified exact expression of SEP for cross QAM in AWGN channel from M$\times$ N rectangular QAM and its usefulness in nakagami-m fading channel. AEU-Int. J. Electron. Commun. **74**, 63–74 (2017)
209. P.K. Singya, P. Shaik, N. Kumar, V. Bhatia, M.-S. Alouini, A survey on higher-order QAM constellations: technical challenges, recent advances, and future trends. IEEE Open J. Commun. Society **2**, 617–655 (2021)
210. P.K. Singya, N. Kumar, V. Bhatia, M.-S. Alouini, On performance of hexagonal, cross, and rectangular QAM for multi-relay systems. IEEE Access **7**, 60 602–60 616 (2019)
211. D. Kumar et al., On performance of a SWIPT enabled FD CRN with HIs and imperfect SIC over α-μ fading channel. IEEE Trans. Cog. Commun. Networ. **9**(1), 99–113 (2023)
212. R. Long, Y.-C. Liang, Y. Pei, E.G. Larsson, Active reconfigurable intelligent surface-aided wireless communications. IEEE Trans. Wirel. Commun. **20**(8), 4962–4975 (2021)
213. Y. Li, Y. Li, X. Chu, Y. Ye, H. Zhang, Performance analysis of relay selection in cooperative NOMA networks. IEEE Commun. Lett. **23**(4), 760–763 (2019)
214. Y. Ye, Y. Li, F. Zhou, N. Al-Dhahir, H. Zhang, Power splitting-based SWIPT with dual-hop DF relaying in the presence of a direct link. IEEE Syst. J. **13**(2), 1316–1319 (2019)
215. Y. Li, T. Li, Y. Li, Q. Ni, C. Zarakovitis, Sum-rate maximization based relay selection for cooperative NOMA over nakagami-*m* fading. IEEE Trans. Veh. Technol. **69**(11), 13 985–13 989 (2020)
216. H.A. David, H.N. Nagaraja, *Order Statistics*, 3rd edn. (Wiley, New York, NY, USA, 2003)
217. K.A. Hamdi, A useful lemma for capacity analysis of fading interference channels. IEEE Trans. Commun. **58**(2), 411–416 (2010)
218. P. Swami, M.K. Mishra, V. Bhatia, T. Ratnarajah, Performance analysis of NOMA enabled hybrid network with limited feedback. IEEE Trans. Veh. Technol. **69**(4), 4516–4521 (2020)

MIX
Papier aus verantwortungsvollen Quellen
Paper from responsible sources
FSC® C105338

If you have any concerns about our products,
you can contact us on
ProductSafety@springernature.com

In case Publisher is established outside the EU,
the EU authorized representative is:
Springer Nature Customer Service Center GmbH
Europaplatz 3, 69115 Heidelberg, Germany

Printed by Libri Plureos GmbH
in Hamburg, Germany